One-Dimensional Microwave Photonic Crystals

One-Dimensional Microwave Photonic Crystals

New Applications

D. A. Usanov
S. A. Nikitov
A. V. Skripal'
D. V. Ponomarev

Taylor & Francis Group
Boca Raton London New York

CISP

CRC Press is an imprint of the
Taylor & Francis Group, an **Informa** business

Translated from Russian by V.E. Riecansky

CRC Press
Taylor & Francis Group
6000 Broken Sound Parkway NW, Suite 300
Boca Raton, FL 33487-2742

First issued in paperback 2021

© 2019 by CISP
CRC Press is an imprint of Taylor & Francis Group, an Informa business

No claim to original U.S. Government works

ISBN 13: 978-1-03-224010-7 (pbk)
ISBN 13: 978-0-367-22656-5 (hbk)

Visit the Taylor & Francis Web site at
http://www.taylorandfrancis.com

and the CRC Press Web site at
http://www.crcpress.com

Contents

Contents

Introduction

The idea of the possibility of the decay of the continuous energy spectrum of electrons into a set of alternating resolved and forbidden bands in the direction of electron motion and wave propagation in the presence of spatial periodicity of the deformation field was first advanced by L.V. Keldysh [1]. Periodic semiconductor structures with the predetermined parameters of layers were called semiconductor superlattices. The wide interest in the problem of their creation appeared after the publication in 1970 of the articles of Esaka and Tsu, who proposed to make such structures by changing the doping or composition of the layers [2]. The periods in such structures had values from 5 to 20 nm. The number of layers reached several hundred.

The author of [3] cited the definition of photonic crystals as materials whose crystal lattice has a periodic dielectric constant leading to the appearance of 'forbidden. frequencies forming the so-called photonic forbidden band. Yablonovich [4] and John [5] proposed creating structures with a photonic band gap, which can be considered as an optical analog of the band gap in semiconductors. The forbidden band in this case is a frequency region in which the propagation of light waves in the inner part of the crystal is impossible. The type of defect or violation of periodicity in this case can be different. Such structures have to be created artificially in contrast to natural crystals. In this case, the size of the basic unit element of a photonic crystal should be comparable with the wavelength of light. For devices used in optoelectronics, the working wavelength is approximately equal to 1.5 μm [3]. The fabrication of such structures involves the use of electron-beam and X-ray lithography [6].

As the advantage of such photonic crystals, the author of [3] notes the possibility of an exact description of their properties that coincides with experiment, in contrast to the superlattices.

The structures with spatial periodicity of elements were also used in the microwave range to reduce the phase velocity of the wave in comparison with the speed of light in special waveguides, called slowing-down systems [7]. The authors of [7] called them 'like artificial crystals, the cells of which are large'.

The slowing-down systems are used in various types of vacuum microwave electronics. The specificity of the slowing-down systems is the choice of their basic elements from metals and the need to take into account in the design the possibility of passing the electron beam interacting with the field. Examples of periodic structures used in slowing-down systems are 'meander', 'counter pins', systems with alternating diaphragms, etc.

In the microwave range, the photonic crystal can be realized both with the help of waveguides with dielectric filling [8, 9], and with the use of plane transmission lines with a periodically changing strip structure [10]. There are examples of the creation of photonic crystals in the optical, infrared, ultraviolet and microwave range. For the microwave range, creating a photonic crystal is the simplest. We note that in the theoretical description of the properties of such structures, unlike, for example, the superlattices, it is not necessary to take into account the properties of transition layers, quantum-mechanical size effects, and the specificity of technological processes. This makes it possible to more accurately consider the properties of the photonic crystals associated with periodicity, and, in particular, to use the results of a theoretical description to measure the parameters of the layers entering the crystal as a result of solving the corresponding inverse problem.

Materials with the properties of photonic crystals are also known in nature. They include, for example, noble opal [11], spicules of natural biomineral crystals – basal spicules of glass sea sponges [12]. The time-varying forbidden band for the frequency region in the vicinity of 6 GHz was observed in a solution with a chemical self-oscillating Briggs–Rauscher reaction characterized by the presence of periodically located regions with different permittivity values [13].

The first two chapters describe the properties of one-dimensional photonic crystals based on rectangular waveguides and plane transmission lines. The third is devoted to the possibility of controlling the characteristics of the microwave photonic crystals with the help of electric and magnetic fields. In the fourth chapter, the possibilities of using the microwave photonic crystals are considered, including examples of new areas of their application, namely, for measuring the parameters of semiconductor layered structures containing nanometer metal layers and as microwave-matched loads.

References

1. Keldysh L.V., On the influence of ultrasound on the electronic spectrum of a crystal, Fiz. Tverd. Tela. 1962. V. 4. No. 8. P.2.265–2267.

2. Esaki L., Tsu R., Superlattice and Negative Differential Conductivity in Semiconductors. IBM Journal of Research and Development. 1970. Vol.14, Issue 1. P.61–65.

3. Joannopulos J., Villeneuve P.R., Fan S., Photonic crystals: putting on new twist on light. Nature. 1997. V. 386. P. 143–146.

4. Yablonovitch E., Inhibited spontaneous emission in solid-state physics and electronics. Phys. Rev. Lett. 1987. V. 58. P. 2059–2063.

5. John S., Strong localization of photons in certain disordered dielectric superlattices Phys. Rev. Lett. 1987. V. 58. P. 2486–2491.

6. Zaitsev D.F., Nanophotonics and its application. Moscow. Akteon. 2012.

7. Silin RA., Sazonov V.P., Slowing-down systems. Moscow, Sov. Radio, 1966.

8. Tae-Yeoul, Kai Chang, Uniplanar one-dimensional photonic-bandgap structures and resonators. IEEE Transactions on Microwave Theory and Techniques. 2001. Vol. 49, N. 3. P. 549–553.

9. Usanov D., Skripal' Al., Abramov A., Bogolubov A., Skvortsov V., Merdanov M., Measurement of the Metal Nanometer Layer Parameters on Dielectric Substrates using Photonic Crystals based on the Waveguide Structures with Controlled Irregularity in the Microwave Band. Proc. of 37th European Microwave Conference. Munich, Germany. 8–12th October 2007. P. 198–201.

10. Gulyaev Yu.V., Nikitov S.A., Photonic and magnetophotonic crystals – a new medium for information transfer. Radiotekhnika, 2003.No. 8. P. 26–30.

11. Samoilovich M.I., Tsvetkov M.Yu., Rare-earth opal nanocomposites for nanophotonics. Nano- i mikrosistemnaya tekhnika. 2006. No. 10. P.8–14.

12. Kulchin Yu.N., Bagaev S.N., Bukin O.A., Voznesensky S.S., Drozdov A.L., Zinin Yu.A., Nagorny I.G., Pestryakov E.V., Trunov V.I., Photonic crystals on the basis of natural biominerals of oceanic origin. Pis'ma Zh. Tekh. Fiz. 2008. V.34, No. 15. P.1–7.

13. Usanov D.A., Rytik A.P., Properties of a photonic crystal formed by a solution with a self-oscillating Briggs-Rauscher reaction. Pis'ma Zh. Tekh. Fiz. 2016. Vol. 42, No. 12. P.45–50.

One-dimensional microwave photonic crystals based on rectangular waveguides

The results of a theoretical analysis of the characteristics of one-dimensional microwave photonic crystals made on the basis of a rectangular waveguide and their experimental investigation are given, for example, in [1]. The one-dimensionality of the crystal means that the dielectric structure that fills the waveguide has periodicity in dielectric permittivity in one direction (along the wave propagation direction, along the Z axis). The authors of [1] studied structures in the form of alternating layers with high dielectric permittivity h and low l (hl-pairs) and layers of the h_1lh_2 type (h_1 and h_2 are dielectrics with different permittivity). In the theoretical description, the scattering matrix method was used. Schematically, the image of the h–l microwave photonic crystal is shown in Fig.1.1. The arrangement of the dielectric layers in the waveguide can be characterized by the following relationships:

$$\varepsilon(x,y,z+d)=(x,y,z),\ 0\le x\le a,\ 0\le y\le b, z<\infty, \tag{1.1}$$

$$\varepsilon(x,y,z+d)=\begin{cases} \varepsilon_1=\varepsilon_0\varepsilon_{r1} \text{ for}-w\le z\le 0 \\ \varepsilon_2=\varepsilon_0\varepsilon_{r2} \text{ for } 0\le z\le v \end{cases}. \tag{1.2}$$

Here ε_{r1} and ε_{r2} are the relative dielectric permittivities of the two materials, ε_0 is the dielectric constant of the vacuum, $d = v + w$.

Using the continuity conditions on the surfaces $z = 0$, $z = v$ and $z = -w$ for the tangential field components, we obtain an equation for finding the wave number γ:

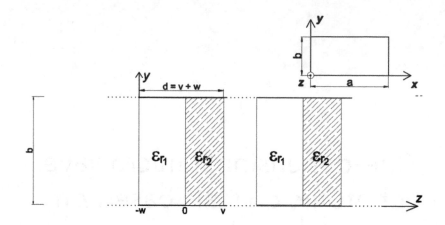

Fig. 1.1. Schematic representation of the periodic change in dielectric permittivity of layers in accordance with the Kronig–Penney model for an ideal structure with an electromagnetic band gap in a rectangular waveguide [1].

$$\gamma = \frac{1}{v+w}\arccos(L_{TE_{10}^z}),\qquad(1.3)$$

where $L_{TE_{10}^z} = \cos(\beta_1 w)\cos(\beta_2 v) - \dfrac{\beta_1^2 + \beta_2^2}{2\beta_1\beta_2}\sin(\beta_1 w)\sin(\beta_2 v)$,
$\beta_{1,2}^2 = \omega^2\mu_0\varepsilon_{1,2} + k^2,\ k^2 = (\pi/a)^2.$

The reflection and transmission coefficients were proposed by the authors of [1] using the generalized scattering matrix, which is obtained as a result of the application of the continuity conditions on the media boundaries. The results of measurements of the transmission coefficient (S_{21}) for microwave photonic crystals consisting of 10 and 20 separate two-layer elements are shown in [1] (Fig. 1.2). The same figure shows the corresponding calculated dependences. The results of the calculation are in good agreement with experiment. Figures 1.2 *c, d* show the results of the calculated and experimental dependences for the value *d* = 9 mm, at which transmission bands are observed along with the stop band, The difference between experiment and theory is attributed by the authors of [1] to the imperfection of the walls of the waveguide.

Figure 1.3 shows the theoretical and experimental results for structures with the parameters shown in Fig. 1.2 *c, d*, with a change in the width of the dielectric layer located in the centre of the structure from 7 to 28 mm. The authors of [1] note that there is no noticeable difference in the results if the defect is introduced into the *n*-th element at $6 \le n \le 13$. Figure 1.4 shows the results

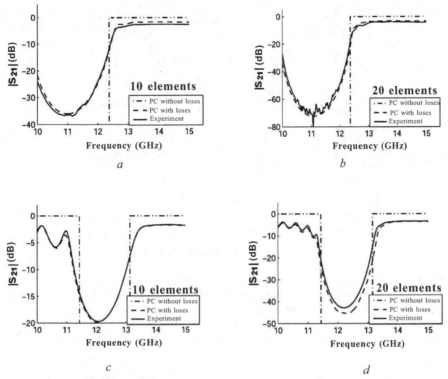

Fig.1.2. Comparison of the calculated and measured frequency dependences of the transmission coefficient (S_{21}) for two one-dimensional structures with an electromagnetic band gap, consisting of 10 and 20 separate two-layer elements, with parameters $v = 7$ mm, $\varepsilon_{r1} = 1$ and $\varepsilon_{r2} = 2.625(1 - j9 \times 10^{-3})$. In the figures (*a*) and (*b*), $d = 11$ mm, in (*c*) and (*d*) $d = 9$ mm [1].

of calculations and measurements for structures with two-layer and three-layer elements. It is noted that the third element adds a new degree of freedom in the design of microwave photonic crystals with the required properties.

To calculate the reflection coefficients R and the transmission coefficient D of an electromagnetic wave upon its normal incidence on a layered structure consisting of N layers, a wave transfer matrix between regions with different values of propagation constants of the wave can be used, just as it was done in [2–4]:

$$\mathbf{T}(z_{j,j+1}) = \begin{bmatrix} \dfrac{\gamma_{j+1} + \gamma_j}{2\gamma_{j+1}} e^{(\gamma_{j+1} - \gamma_j)z_{j,j+1}} & \dfrac{\gamma_{j+1} - \gamma_j}{2\gamma_{j+1}} e^{(\gamma_{j+1} + \gamma_j)z_{j,j+1}} \\[3mm] \dfrac{\gamma_{j+1} - \gamma_j}{2\gamma_{j+1}} e^{-(\gamma_{j+1} + \gamma_j)z_{j,j+1}} & \dfrac{\gamma_{j+1} + \gamma_j}{2\gamma_{j+1}} e^{-(\gamma_{j+1} - \gamma_j)z_{j,j+1}} \end{bmatrix}, \quad (1.4)$$

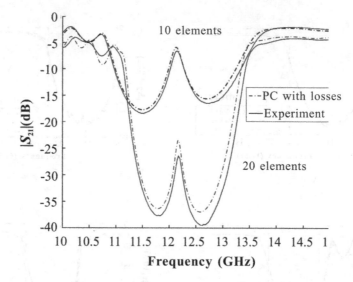

Fig.1.3. Influence of the disturbance in the structure with the electromagnetic band gap, presented in Fig. 1.2 *b*. The dimensions of the perturbed two-layer cell element are $v = 28$ mm and $w = 2$ mm. For a structure with an electromagnetic band gap consisting of 10 separate two-layer elements, the disturbance is located in the fifth position, for a structure consisting of 20 two-layer elements – in the ninth [1]

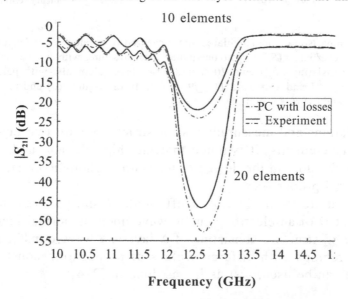

Fig.1.4. Calculated and measured frequency dependences of the transmission coefficient (S_{21}) for two one-dimensional structures with an electromagnetic band gap (waveguide type R120), consisting of 10 and 20 separate three-layer elements, with parameters $w = 7$ mm, $v = 2$ mm, $u = 7$ mm, $\varepsilon_{r2} = 2.625$ $(1 - j9 \times 10^{-3})$, $\varepsilon_{r2} = 1$ and $\varepsilon_{r3} = 3.025$ $(1 - j7.9 \times 10^{-3})$. Input and output waveguides are empty [1].

which connects the coefficients A_j, B_j and A_{j+1}, B_{j+1}, which determine the amplitudes of the incident and reflected waves on both sides of the boundary $z_{j,j+1}$, by the relation:

$$\begin{bmatrix} A_{j+1} \\ B_{j+1} \end{bmatrix} = \mathbf{T}(z_{j,j+1}) \cdot \begin{bmatrix} A_j \\ B_j \end{bmatrix}. \tag{1.5}$$

The coefficients A_{N+1} and B_0, which determine the amplitudes of the wave transmitted through the multilayer structure (Fig. 1.5), and the wave reflected from it, are related to the coefficient A_0 determining the amplitude of the incident wave by the following relation:

$$\begin{bmatrix} A_{N+1} \\ 0 \end{bmatrix} = \mathbf{T}_N \cdot \begin{bmatrix} A_0 \\ B_0 \end{bmatrix}, \tag{1.6}$$

where

$$\mathbf{T}_N = \begin{bmatrix} \mathbf{T}_N[1,1] & \mathbf{T}_N[1,2] \\ \mathbf{T}_N[2,1] & \mathbf{T}_N[2,2] \end{bmatrix} = \prod_{j=N}^{0} \mathbf{T}_{j,(j+1)} =$$

$$= \mathbf{T}(z_{N,N+1}) \cdot \mathbf{T}(z_{N-1,N}) \dots \mathbf{T}(z_{1,2}) \cdot \mathbf{T}(z_{0,1}) \tag{1.7}$$

is the transfer matrix of a layered structure consisting of N layers.

The reflection coefficients $R = \dfrac{B_0}{A_0}$ and the transmission coefficients $D = \dfrac{A_{N+1}}{A_0}$ of an electromagnetic wave interacting with a layered structure are determined by the following relationships:

$$R = \frac{\mathbf{T}_N[2,1]}{\mathbf{T}_N[2,2]}, \tag{1.8}$$

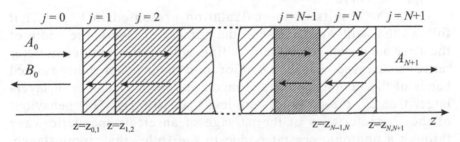

Fig.1.5. Structure consisting of N layers [2].

$$D = \frac{\mathbf{T}_N[1,1] \cdot \mathbf{T}_N[2,2] - \mathbf{T}_N[1,2] \cdot \mathbf{T}_N[2,1]}{\mathbf{T}_N[2,2]}. \tag{1.9}$$

R and D were calculated using the transmission matrices of the wave between the regions with values of the propagation constant of the electromagnetic wave γ_j and γ_{j+1}.

The results of a theoretical and experimental study of the resonant features that arise in the allowed and forbidden bands of an ultrahigh-frequency photonic crystal in violation of the periodicity are given in [5].

The appearance of a resonance feature in a photonic band gap, called an impurity or defect resonant mode, in the presence of a defect in the structure of a photonic crystal that violates its periodicity, for example, in the form of a change in the electrophysical parameters of one or several elements, was considered in Refs. [6, 7]. The change in the parameters of the perturbations created by the microwave photon crystal, including the use of temperature [8], electric [9], magnetic fields [10], allows us to control the frequency positions of the impurity resonant mode in the forbidden band of a photonic crystal. As a waveguide photonic crystal we used a section of a waveguide with a structure that is a section of a waveguide with periodically alternating layers of two types of dielectrics with different values of thickness and dielectric constant. The dimensions and materials of the layers were chosen in such a way that in the frequency range 8–12 GHz two resolved and one forbidden band were observed for the propagation of electromagnetic waves. The parameters of the first and second layers of the photonic crystal were the same. The results of calculating the square of the modulus of the transmission coefficient D of an electromagnetic wave using the above relationships for the 11-layer structure of a photonic crystal without perturbations during propagation of a H_{10} wave are shown in Fig. 1.6 (curve *1*).

From the results of the calculation presented in Fig. 1.7, it follows that with an increase in the number of layers, the width of the first band gap completely falling within the three-centimeter range is observed, and the width of both the left and right resolved bands of this frequency band is increased. With a number of layers larger than 27, these changes are less than 10 MHz. This behaviour of the characteristics of the passage of an electromagnetic wave through a photonic crystal is due to the following circumstance.

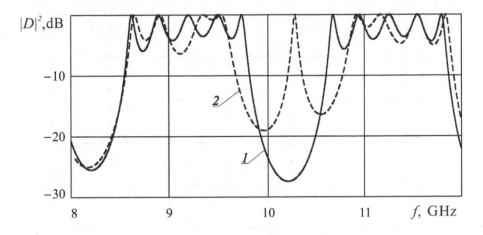

Fig. 1.6. Results of calculation of the squared modulus of the transmission coefficient of the electromagnetic wave through the 11-layer structure without perturbation (curve *1*) and with the perturbed central layer (curve *2*) [5].

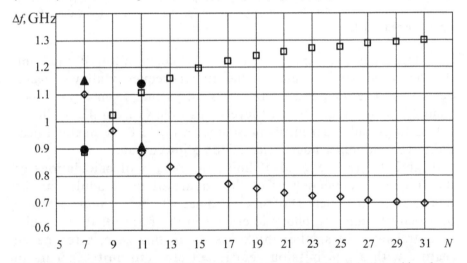

Fig. 1.7. Dependences of the width of the forbidden and left-hand allowed bands on the number of layers of a photonic crystal without disturbance: □, • – theoretical and experimental values of the width of the left allowed band, ◊, ▲ – theoretical and experimental values of the band gap [5].

The allowed band has a 'cut' frequency response and consists of a set of resonances, the number of which is determined by the number of identical elements that make up its composition, so with increasing number of layers of the photonic crystal, the number of resonances determining the width of the allowed band increases and, consequently, its width increases. At the same time, the width of the forbidden band, defined as the frequency gap between the

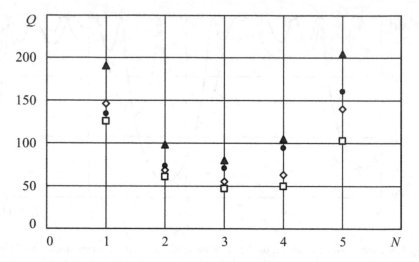

Fig.1.8. The goodness of the resonances forming the left and right allowed bands for the 11-layer structure: •, ▲ – the experimental Q values for left and right allowed resonances, □, ◊ – theoretical Q values for the resonances+ of the left and right allowed bands [5].

high-frequency resonance of the low-frequency allowed band and the low-frequency resonance of the higher-frequency allowed band, decreases. The goodness of the resonances that form the allowed band of the so-called 'allowed levels', as follows from the results of calculations and experiments presented in Fig. 1.8, is maximal near the lower and upper boundaries of the allowed band.

Figure 1.6 (curve *2*) also shows the results of calculations of the frequency dependences of the square of the modulus of the transmission coefficient D of an electromagnetic wave in the presence of a perturbation in a photonic crystal in the form of, for example, a central layer of smaller thickness d_6. In this case, a resonance feature with a transmission coefficient close to unity appears in the band gap. An impurity resonant mode whose location can be controlled by varying the thickness of the dielectric layer. A defect in a photonic crystal, as shown in [5], also leads to a change in the number and position on the frequency axis of the resonances that form the allowed band. When a violation in the photonic crystal is created in the form of a central layer of a smaller thickness, the number of resonances in the allowed band decreases by one in comparison with the photonic crystal without the perturbation, and the high-frequency resonance peak of the allowed band adjoining the low-frequency side to the forbidden band in which an impurity resonant mode has arisen, is shifted to the low-frequency region to a position close to the position of the next resonance peak (i.e., the high-frequency resonance peak of the allowed band that is adjacent to the low-frequency side to the forbidden zone, in which an impurity resonant mode arose). In this case, the low-frequency resonant peak of the allowed band adjoining the high-frequency side to the forbidden band in which an impurity resonant mode arises is shifted to the high-frequency region at a position close to the position of the next resonant peak (i.e. in this case the low-frequency resonance peak of the allowed band disappears, adjoining from the high-frequency side to the forbidden band, in which an impurity resonant mode arose).

Thus, the width of the frequency region between the last high-frequency resonance peak of the allowed band adjoining the low-frequency side to the forbidden band is increased, and the first low-frequency resonance peak of the allowed band adjacent to the high frequency side to the forbidden band. This situation can be interpreted as an increase in the width of the forbidden band containing an impurity mode (see Fig. 1.6, curve *2*). Since the allowed band, as noted earlier, consists of a set of resonances, the number of which is determined by the number of identical elements that make up its composition, the introduction into the photonic crystal of a violation, for example, as a central layer of smaller thickness, leads to a decrease in the number of identical elements of the photonic crystal, consequently, a decrease in the number of resonances composing the allowed band, and, thus, to a significant increase in the forbidden width, especially for a small number of periods of a photonic crystal.

Since the number of resonance peaks in the allowed bands increases with the number of layers of the photonic crystal, and the dimensions of the forbidden and allowed bands, as noted above, vary relatively little, in this case the size of the frequency region between the resonance peaks that determines the width of the forbidden band containing the impurity mode tends in the limiting case (when the number of layers tends to infinity) to the width of the forbidden band of a photonic crystal without disturbances (Fig. 1.9). It should be noted that the frequency position of the resonance of the impurity mode within the forbidden band, depending on the parameters of the inhomogeneity, in a rather wide range of their variation affects the magnitude of the forbidden band insignificantly.

In the experiments we used one-dimensional waveguide photonic crystals consisting of seven and eleven layers completely filling a

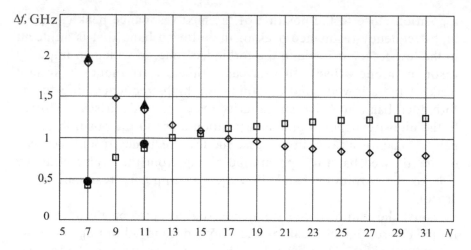

Fig. 1.9. Dependence of the width of the forbidden and left-hand allowed bands on the number N of layers of a photonic crystal with a perturbation: □, ● – theoretical and experimental values of the width of the left-allowed band, ◊, ▲ – theoretical and experimental values of the band gap [5].

rectangular waveguide of a three-centimeter range of wavelengths along the cross section. Odd layers were made of ceramic (Al_2O_3, $\varepsilon = 9.6$), even – from Teflon ($\varepsilon = 2.1$). The length of odd segments was 1 mm, even 44 mm. The perturbation was created by changing the length of the central layer, which led to the appearance of an impurity resonant mode in the forbidden band of a photonic crystal. The length of the central perturbed (Teflon) layer was chosen equal to 14 mm. The frequency dependences of the reflection and transmission coefficients of microwave radiation interacting with a photonic crystal were measured with the Agilent PNA-L Network Analyzer N5230A vector analyzer (Fig. 1.10 *a*, *b*).

As follows from the results of the experiment presented in Fig. 1.10 *a*, *b*, in the frequency range from 8–12 GHz, two allowed and one forbidden bands were observed for the propagation of electromagnetic waves in a photonic crystal (curves *1*), and when the length of the central layer of a photonic crystal in the forbidden band, an impurity resonant mode appeared (curves *2*). The values of the width of the forbidden and allowed bands for the 7- and 11-layer photonic crystal were determined experimentally in the absence (see Fig. 1.7) and the presence of a violation (see Fig. 1.9) in the central layer. Comparison of results of calculation and experiment testifies to their good quantitative coincidence. From the experimental results presented in Fig. 1.10 *a*, *b*, it follows that, as

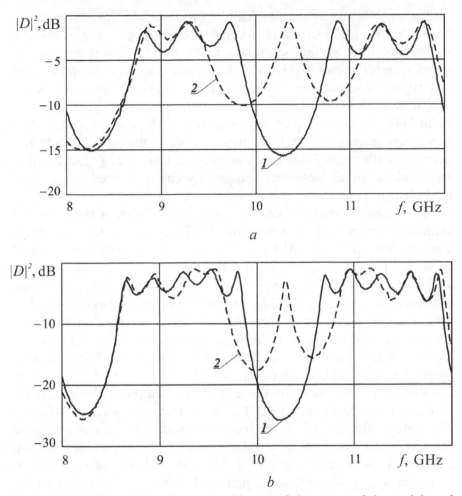

Fig.1 .10. Experimental frequency dependences of the square of the modulus of transmission coefficients of microwave radiation interacting with a 7-layer (*a*) and 11-layer (*b*) photonic crystal without perturbation (curve *1*) and with a perturbation of the length of the central layer (curve *2*). The length of the perturbed layer is 14 mm [5].

was theoretically shown, the existence of an impurity resonant mode in the forbidden band of a photonic crystal leads to an increase in its width. In this case, an increase in the number of layers of a photonic crystal leads to a decrease in the width of the forbidden band, in the presence of an impurity resonant mode. The *Q*-factor of the resonances forming the allowed band was determined experimentally, the so-called 'allowed levels' (see Fig. 1.8). As follows from the results of measurements, the resonance quality factor is maximal near the lower and upper boundaries of the allowed bands.

The established regularities may be of interest in the development, based on photonic crystals, of microwave circuit elements and highly sensitive methods for measuring the electrical parameters of nanometer metal, dielectric and semiconductor materials and structures, composites used in micro-, nano-, and microwave electronics.

In [11], the results of an investigation of the characteristics of a waveguide microwave photonic crystal made in the form of dielectric matrices with air inclusions are presented. Ceramics (Al_2O_3) with a large number of air inclusions and polystyrene were used as materials of dielectric layers,

A waveguide photonic crystal consisting of eleven layers in the frequency range 8–12 GHz was studied (Fig. 1.11 *a*). The odd layers were made of ceramics (Al_2O_3, $\varepsilon = 9.6$), the even layers were made of expanded polystyrene ($\varepsilon = 1.05$). The thickness of the odd segments $d_{Al_2O_3} = 1.0 \text{ mm}$, even $d_{foam} = 13.0 \text{ mm}$. The layers completely filled the cross section of the waveguide. In ceramic layers, a large number of air inclusions are created in the form of square through holes, which form a periodic structure in the plane of the layer (Fig. 1.11 *b*).

On the basis of numerical modelling using the finite element method in the CAD ANSYS HFSS, the influence of the volume fraction of air inclusions on the amplitude–frequency characteristics of the transmission coefficient of a photonic crystal was investigated. The volume fraction of the air inclusions was regulated by changing the size of the apertures a_{hole} in the ceramic plates at a fixed amount equal to 36 in each of the plates. Simulations were carried out for three sizes a_{hole} of the holes, equal to 0.75 mm, 1.2 mm, 1.65 mm, which corresponds to a volume fraction of air inclusions equal to 8.5%, 23% and 43%, respectively.

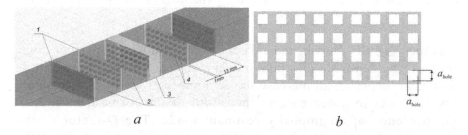

a   b

Fig.1.11. Model of the photonic crystal: *1* – layers of ceramic with square through holes, *2* – layers of foam, *3* – perturbation in the form of a layer of expanded plastic foam of a changed thickness, *4* – square holes (*a*). A layer of ceramics with square through holes (*b*) [11].

The violation of the periodicity of the photon structure was created by a change in the thickness d_{6foam} of the central (sixth) layer of the foam. The thickness of the sixth modified (broken) layer of the foam was 2.25 mm, 2.75 mm, 3.49 mm and 6.0 mm.

The parameters of the photonic crystal were chosen in such a way that when the proportion of air volume inclusions in the ceramic layers varies from 0 to 43%, the resulting photonic forbidden bands overlap a significant part of the used 8–12 GHz frequency range, and the transmission peak due to the created perturbation of the photon structure remained within the forbidden band.

As follows from the results of the numerical calculation of the amplitude–frequency characteristics of the amplitude–frequency response of the transmission coefficient D of a photonic crystal, shown in Fig. 1.12, in the absence of a violation of the periodicity of the photonic structure, an increase in the volume fraction of air inclusions in the ceramic layers leads to a shift in the forbidden band of the photonic crystal toward more short wavelengths and to a decrease in its depth. The results of calculating the amplitude-frequency characteristics D of a photonic crystal by the finite element method in the presence of a violation of the periodicity of the photonic structure in the form of a central layer of foam plastic with different thickness are shown in Fig. 1.1.13–1.16 (solid curves).

The presence of a perturbation of the periodicity of the photon structure in the form of a change in the thickness d_{6foam} of the central (sixth) layer of foam, resulted in the appearance of a transmission peak in the forbidden band of the photonic crystal, the position of which is determined by the size of this perturbation. The change in the thickness d_{6foam} of the impaired foam layer in the range from 2.25 to 6.0 mm in the absence of air inclusions caused a change in the position of the transmission peak from 10.88 to 8.89 GHz, while with a volume fraction of inclusions equal to 43%, the position of the transmission peak varied from 10.62 to 12.0 GHz. It also follows from the obtained results that for different values of the thickness of the broken sixth layer of the structure, an increase in the volume fraction of air inclusions leads to a shift in the forbidden band of the photonic crystal toward shorter wavelengths and to a decrease in its depth. The characteristics of the photonic crystal created in accordance with the model described above were investigated experimentally. Measurement of the amplitude-frequency characteristic of the transmittance of the investigated photonic crystal in the three-centimeter wavelength range was carried out using

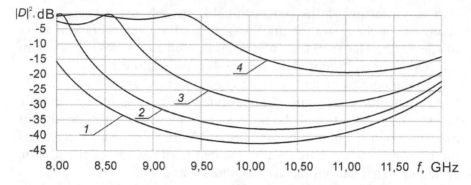

Fig. 1.12. Calculated frequency response of the transmittance of a photonic crystal without disturbance at various volume fractions of air inclusions. The volume fraction of inclusions: *1* – 0%, *2* – 8.5% *3* – 23.0% *4* – 43.0% [11].

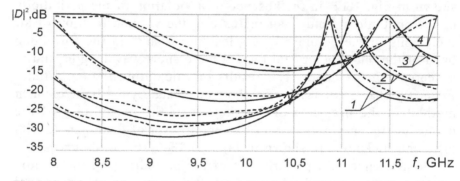

Fig. 1.13. Calculated (continuous curves) and experimental (discrete curves) amplitude–frequency characteristics of the transmittance of a photonic crystal for various volume fractions of air inclusions with a perturbed sixth layer of foam plastic with a thickness of 2.25 mm. The volume fraction of inclusions: *1* – 0%, *2* – 8.5%, *3* – 23.0%, *4* – 43.0% [11].

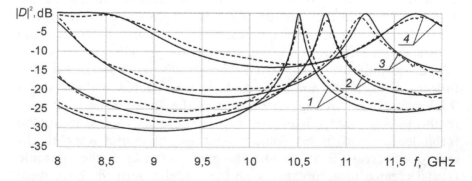

Fig. 1.14. Calculated (continuous curves) and experimental (discrete curves) amplitude–frequency characteristics of the transmittance of a photonic crystal for various volume fractions of air inclusions with a perturbed sixth layer of 2.75 mm thick foam. The volume fraction of inclusions: *1* – 0%, *2* – 8.5%, *3* – 23.0%, *4* – 43.0% [11].

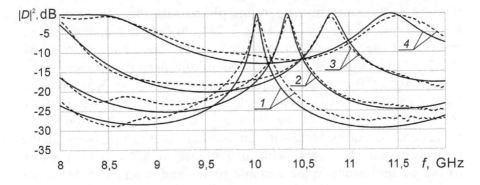

Fig.1. 15. Calculated (continuous curves) and experimental (discrete curves) amplitude–frequency characteristics of the transmittance of a photonic crystal for various volume fractions of air inclusions with a perturbed sixth layer of foam 3.49 mm thick. The volume fraction of inclusions: *1* – 0%, *2* – 8.5%, *3* – 23.0%, *4* – 43.0% [11].

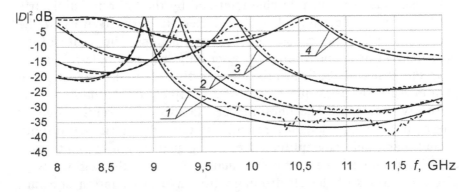

Fig. 1.16. Calculated (continuous curves) and experimental (discrete curves) amplitude–frequency characteristics of the transmittance of a photonic crystal for various volume fractions of air inclusions with a disrupted sixth layer of 6.0 mm thick foam. The volume fraction of inclusions: *1* – 0%, *2* – 8.5%, *3* – 23.0%, *4* – 43.0% [11].

the Agilent PNA-L Network Analyzer N5230A vector analyzer. The results of experimental studies of the amplitude–frequency characteristics of the transmittance of a photonic crystal without disturbance and at different thicknesses d_{6foam} of the perturbed layer of foam is shown in Fig. 1.13 – 1.16 (discrete curves). Comparison of the results of the calculation and the experiment indicates their good agreement. Figure 1.17 shows the results of computer simulation (solid curves) and experimental study (triangles) of the frequency position of the transmission peak in the forbidden band of a photonic crystal from the size a_{hole} through square holes in ceramic plates, acting as air inclusions, at different thicknesses of the broken layer of foam. It also follows from the data obtained that the increase in the size of the holes in the ceramic plates with the thickness of the broken layer of the foam plastic 2.25 mm led to a change in the position of the transmission peak in the forbidden band from 10.88 to 12.0 GHz, while at the thickness of the perturbed layer of the foam plastic 6.0 mm the position of the transmission peak changed from 8.89 to 10.62 GHz. Thus, the sensitivity of the position of the transmission window to the change in the volume fraction of air inclusions increases both with an increase in the thickness of the perturbed foam layer and with an increase in the volume fraction of the air inclusions in the ceramic plates and reaches 40 MHz/%.

The layers of the investigated photonic crystals containing a large number of air inclusions can be considered as composite materials, which are dielectric matrices based on the ceramics with filler in the form of air inclusions. It is known that the dielectric properties of composite materials can be characterized by the effective dielectric permittivity ε_{ef}, determined by the dielectric permittivities of the matrix ε_1 and the filler ε_2 and their volume fractions. The results of the investigation of the possibility of describing the amplitude-frequency characteristics of the transmission coefficient of the investigated photonic crystal using the model of an 'effective' medium [12–14] were given in [11]. The photonic crystal was then represented as alternating homogeneous layers with effective dielectric permittivity ε_{ef} and foam, and its amplitude–frequency characteristics (AFC) was calculated using a wave transfer matrix between the regions with different values of the electromagnetic wave propagation constant determined by the effective permittivity of the ceramic layers with air inclusions and the dielectric permittivity of the foam. To determine the effective dielectric permittivity ε_{ef}, it is necessary to solve the

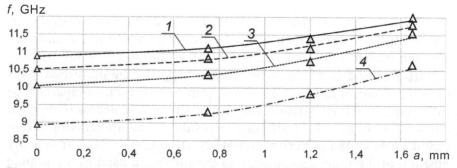

Fig.1.17. Dependence of the position of the transmission peak in the forbidden band of a photonic crystal on the size of through square holes in ceramic plates. The thickness of the perturbed layer is d_{6foam}, mm: *1* – 2.25, *2* – 2.75, *3* – 3.49, *4* – 6.0 [11].

inverse problem [3, 15]. According to the frequency dependences of the transmittance of a photonic crystal consisting of periodically alternating layers of ceramics with air inclusions and foam, the inverse problem was solved using the method of least squares. When implementing this method, it is necessary to determined the value of the parameter ε_{ef}, for which the sum $S(\varepsilon_{ef})$ of the squares of the differences in the transmission coefficients D_{phc} of a photonic crystal consisting of periodically alternating layers of ceramics with air inclusions and foam, and transmittance $D(\varepsilon_{ef}, f_n)$ of a photonic crystal consisting of periodically alternating homogeneous layers with an effective dielectric permittivity ε_{ef} and a foam,

$$S(\varepsilon_{ef}) = \sum_n \left(\left| D_{phc} \right|^2 - \left| D(\varepsilon_{ef}, f_n) \right|^2 \right)^2 \tag{1.10}$$

becomes minimal.

The desired value of the effective dielectric permittivity ε_{ef} of the periodically alternating homogeneous layers is determined by the numerical method as a result of solving equation

$$\frac{\partial S(\varepsilon_{ef})}{\partial \varepsilon_{ef}} = \frac{\partial \left(\sum_n \left(\left| D_{phc} \right|^2 - \left| D(\varepsilon_{ef}, f_n) \right|^2 \right)^2 \right) \varepsilon_{ef}}{\partial \varepsilon_{ef}} = 0. \tag{1.11}$$

Figure 1.18 shows the frequency dependences of the square of the modulus of transmittance of a photonic crystal $|D(f)|^2$ with a different proportion of air inclusions x_1 in layers of ceramics with a perturbed sixth layer of foam $d_{6foam} = 6.0$ mm (circles) calculated using the finite element method in CAD HFSS ANSYS, and the frequency dependences of the transmission coefficients of a photonic crystal (curves) calculated at the values of the effective permittivity ε_{ef} found from the solution of Eq. (1.11).

Figure 1.19 shows the dependence of the quantity ε_{ef}, found from the solution of Eq. (1.11), on the volume fraction x_1 of air inclusions (curve *1*). The same figure shows the results of calculating the effective permittivity of a composite ε_{ef} consisting of a ceramic matrix (with dielectric permittivity ε_2) and air inclusions (with a permittivity ε_1) with a volume fraction x_1 of the latter, made using a number of known models of the 'effective' medium, described by the relations:

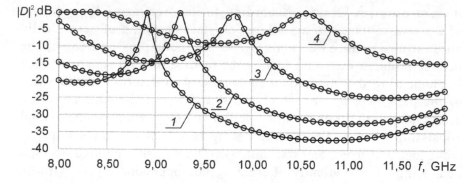

Fig. 1.18. Frequency response of the transmission coefficients of a photonic crystal calculated at the values of the effective dielectric permittivity ε_{ef}, found from the solution of Eq. (1.11), layers of ceramics with a different proportion of air inclusions, x_1, with a perturbed sixth layer of 6.0 mm thickness. ε_{ef}, rel. unit: *1* – 9.6 (x_1 = 0%), *2* – 8.44 (x_1 = 8.5%), *3* – 6.78 (x_1 = 23.0%), *4* – 4.82 (x_1 = 43.0%) [11].

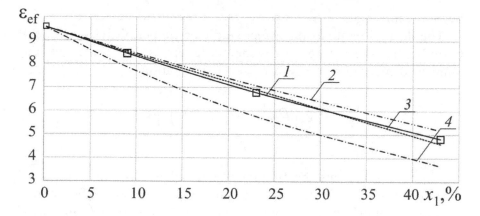

Fig. 1.19. Dependence of the effective permittivity of the layers on the volume fraction of air inclusions. Models of the 'effective' environment: 1 – from the solution of the inverse problem, 2 – Maxwell–Garnett, 3 – Bruggeman, 4 – Lichteneker [11].

Maxwell–Garnett [12]

$$\frac{\varepsilon_{ef} - \varepsilon_2}{\varepsilon_{ef} + 2\varepsilon_2} = x_1 \frac{\varepsilon_1 - \varepsilon_2}{\varepsilon_1 + 2\varepsilon_2} \qquad (1.12)$$

Bruggeman [13]

$$x_1 \frac{\varepsilon_1 - \varepsilon_{ef}}{2\varepsilon_{ef} + \varepsilon_1} = x_2 \frac{\varepsilon_2 - \varepsilon_{ef}}{2\varepsilon_{ef} + \varepsilon_2} = 0 \qquad (1.13)$$

Lichtenecker [14]

$$\log \varepsilon_{ef} = (1 - x_1)\log \varepsilon_2 + x_1 \log \varepsilon_1. \tag{1.14}$$

As follows from the results presented in Fig.1.19, when using the model of the 'effective' medium for calculating the amplitude-frequency characteristics of the transmission coefficient of the investigated photonic crystal in which the photonic crystal was represented as alternating homogeneous layers with an effective dielectric permittivity ε_{ef}, the dependence of ε_{ef} on of the volume fraction of air inclusions, determined using the technique described above, in the range 0–43% coincides with the dependence described by the Maxwell–Garnett relation with an error of 7.47% (see Fig. 1.19, curve *2*), using the Bruggeman equation (see Fig. 19, curve *3*) error is 4.50%, and reaches 24.2% when using the Lichtenecker ratio (see Fig. 1.19, curve *4*).

The analysis of the obtained results indicates the possibility of describing a photonic crystal consisting of periodically alternating layers of ceramics with a large number of air inclusions in the form of square through holes forming in the plane of the layer a periodic structure using the model of an 'effective' medium and the absence of regularities related to the periodicity of the properties of the photonic crystal in the directions orthogonal to the direction of propagation of the electromagnetic wave.

This indicates, in particular, that with such a configuration of a photonic crystal there is no effective conversion of a main type wave to higher-type waves, which is due to the smallness of the transverse dimensions of the inclusions compared to the wide wall of the waveguide for a sufficiently large number of them.

Consequently, to describe such three-dimensional photonic crystals, it is appropriate to use the model of a one-dimensional photonic crystal. In this case, it is possible to use the relations describing the interaction of an electromagnetic wave with a photonic crystal in the form of a transmission matrix for a one-dimensional waveguide photonic crystal, since the band character of the frequency dependence of the transmission coefficient of a photonic crystal, related to the periodicity of the structure of the photonic crystal in the transverse plane for the standard cross section waveguide in the considered range of dielectric permittivities of materials is not manifested.

The established regularities can be used to create dielectric layers with given dielectric permittivity values, which are necessary for the construction of matched broadband microwave loads based on photonic crystals [15, 16].

In the optical range, with the thickness of an individual element approximately equal to half the wavelength, with a large number of elements forming a photonic crystal, the overall size of the device on it remains sufficiently small [17]. A distinctive feature of photonic crystals of the microwave range is the ability to realize the various functions necessary for the operation of microwave circuits [18–21] with a relatively small number of elements that make up a photonic crystal. A small number of elements forming microwave photonic crystals is associated with the need for compactness of the devices created on their basis. At the same time, the frequency characteristics of the devices may not be optimal at the same time.

It is of scientific and practical interest to create multi-element microwave photonic crystals characterized by small dimensions. To solve this kind of problem, it is possible to use as a photonic crystal a structure that excites waves of higher types whose wavelengths are substantially shorter than the wavelength in the waveguide of the basic type, as suggested by the authors of Ref. [22]. Therefore, the dimensions of devices at higher types of waves become significantly smaller than similar devices on the main type of wave. In this regard, they can be called low-dimensional. In [22], the results of investigations of waveguide photonic crystals representing structures from successively alternating dielectric layers–even elements of a photonic crystal and thin metal plates partially overlapping the waveguide cross section–are given by the odd elements of a photonic crystal.

There were gaps between the plates and the wide walls of the waveguide. Each of the plates created the same width along the entire length of the plate. In this case, the gaps between the odd metal plates and the waveguide were created at one of the wide walls of the waveguide, and the gaps between the even metal plates and the waveguide are at the opposite wide wall of the waveguide.

In this case, the gap is the source of waves of higher types, forming in its vicinity the so-called near field. The structure of the investigated low-dimensional waveguide microwave photonic crystal is shown in Fig. 1.20.

The dielectric material used was: expanded polystyrene ($\varepsilon = 1.02$) and fluoroplastic ($\varepsilon = 2.1$). The dielectric layers completely filled

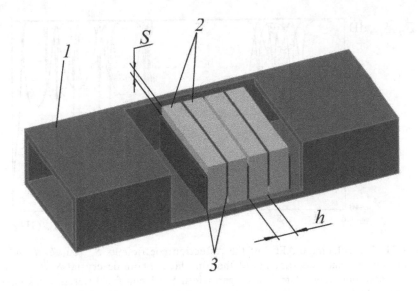

Fig. 1.20. Model of a photonic crystal: *1* – a segment of a rectangular waveguide, *2* – layers of a dielectric, *3* – thin metal plates, *S* – width of a gap, *h* – thickness of a dielectric layer [22].

the cross-section (23×10 mm) of the waveguide. The metal plates 50 μm thick were made of aluminium. The width of the gap *S* did not exceed one tenth of the size of the narrow wall of the waveguide. On the basis of numerical simulation using the finite element method in the ANSYS HFSS CAD, the influence of the thickness of dielectric layers, the gap width and the number of layers of the structure of a photonic crystal on the amplitude–frequency characteristics of the transmission and reflection coefficients of a photonic crystal was investigated.

The parameters of the photonic crystal (PC) were chosen in such a way as to establish the possibility of manifesting the 'band' character of the frequency dependences of the transmittance and reflection coefficients of the PC associated with the periodicity of its structure.

The results of the calculation of the transmission S_{12} and the reflection S_{11} coefficients of a nine-layer photonic crystal consisting of five consecutively alternating metal plates with gaps and 4 dielectric layers at various values of the thickness of the dielectric layers *h* are shown in Fig. 1.21.

From the results obtained, it follows that the amplitude–frequency characteristic of the transmittance S_{12} of the structure under study has a 'band' character. The frequency response of such a photonic crystal consists of characteristic alternating 'allowed' and 'forbidden' bands.

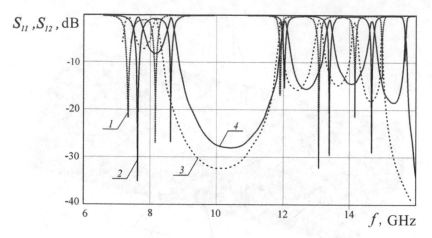

Fig. 1.21. The calculated AFCs of the reflection coefficients S_{11} (curves *1* and *2*) and the transmittance coefficients of the nine-layer photonic crystal S_{12} (curves *3* and *4*) containing metal plates with a gap width $S = 1$ mm for different thicknesses h of the dielectric layers (fluoroplastic, $\varepsilon = 2.1$). Curves *1* and *3* – $h = 3.5$ mm. Curves *2* and *4* – $h = 3$ mm [22].

The 'allowed' bands are formed by a set of resonant transmission peaks with the transmittance values S_{12} in the resonance region to -0.4 dB. The amplitude–frequency characteristic of the reflection coefficient S_{11} of this structure is formed by a set of resonance reflection peaks with values in resonances up to -35 dB. The frequency position of the resonance peaks of the coefficient S_{11} corresponds to the position of the peaks of the coefficient S_{12}.

A decrease in the thickness of the dielectric layers h of the structure, as well as an increase in the width of the gap S, led to a shift in the amplitude–frequency characteristic of the photonic crystal toward shorter wavelengths and to an increase in the width of the 'allowed' band. At the same time, the width and depth of the 'forbidden' band decreased. With a decrease in the thickness of the dielectric layers h from 4 to 3 mm, the width of the 'allowed' band increased by 39%, the depth of the 'forbidden' band decreased by 26%, and its width decreased by 15%. At the same time, an increase in the width of the gap S from 0.8 to 1 mm resulted in an increase in the width of the 'allowed' band by 11%, a decrease in the depth of the 'forbidden' band by 14% and a decrease in its width by 7%.

Investigations of the influence of the number of layers of a photonic crystal on the frequency dependence of the transmission coefficients S_{12} and the reflection coefficients S_{11} have shown that when the number of layers changes, the number of resonances in the 'allowed' band at the frequency response of the transmission coefficients S_{12} and the reflection coefficients S_{11} of the photonic crystal change. It was found that with the number of metal plates in the structure of a photonic crystal equal to m, the number of resonances on the AFC of a photonic crystal is $m-1$.

The photonic crystal created in accordance with the model described above was studied experimentally. Measurement of the frequency dependences of the transmission coefficients S_{12} and the reflection coefficients S_{11} of the investigated photonic crystal was carried out using the Agilent PNA-L Network Analyzer N5230A vector analyzer. Figures 1.22 and 1.23 show the results of experimental studies (solid curves) of the AFC of the transmission coefficients S_{12} and the reflection coefficients S_{11} of the photonic crystals consisting of successively alternating thin metal plates with gaps and dielectric layers of expanded polystyrene ($\varepsilon = 1.02$), with a different number of layers of such a photon crystal. The same figures show the results of the calculation of the frequency response of similar photonic crystals.

Comparison of the results of the calculation and the experiment indicates their good qualitative coincidence.

Thus, the use of metal plates with gaps in the structure of the photonic crystal has made it possible to substantially reduce its

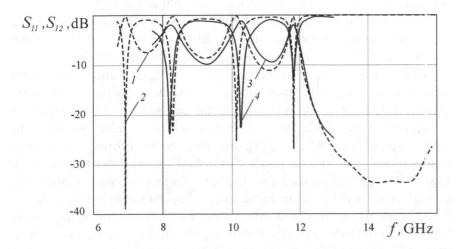

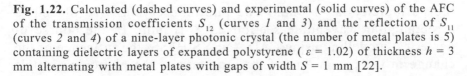

Fig. 1.22. Calculated (dashed curves) and experimental (solid curves) of the AFC of the transmission coefficients S_{12} (curves *1* and *3*) and the reflection of S_{11} (curves *2* and *4*) of a nine-layer photonic crystal (the number of metal plates is 5) containing dielectric layers of expanded polystyrene ($\varepsilon = 1.02$) of thickness $h = 3$ mm alternating with metal plates with gaps of width $S = 1$ mm [22].

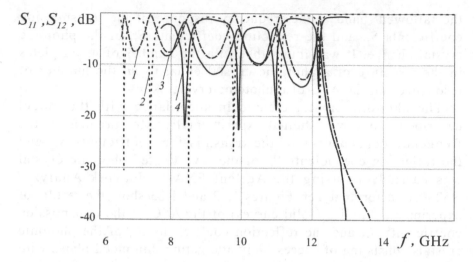

Fig. 1.23. Calculated (dashed curves) and experimental (solid curves) AFC of the transmission coefficients S_{12} (curves *1* and *3*) and reflection coefficients S_{11} (curves *2* and *4*) of a thirteen-layer (number of metal plates 7) photonic crystal containing dielectric layers of expanded polystyrene ($\varepsilon = 1.02$) of thickness $h = 3$ mm alternating with metal plates with gaps of width $S = 1$ mm [22].

longitudinal dimension to 12.25 mm, which is about five times smaller than the longitudinal dimension of a photonic crystal created on elements made of alternating layers of dielectrics with different permittivity and having similar spectral characteristics.

The perturbation of periodicity in a low-dimensional microwave photonic crystal, as well as in the ordinary one, should lead to the appearance of a defect (impurity) mode. A theoretical definition of the conditions for its appearance and their experimental realization were described in [23]. A perturbation of the periodicity of a photonic crystal can be a different size of the central dielectric layer or a modified size of the capacitive gap of the diaphragm adjacent to the perturbed layer. In [23], the results of studies of the low-dimensional photonic crystals described above (see the inset in Fig. 1.24) are presented. The layout of the layers of a photonic crystal is shown in the inset to Fig. 1.25. The diaphragms 50 μm thick were made of aluminium. The thickness of each dielectric layer in a photonic crystal without 'perturbations' was 3 mm. Thus, the total longitudinal dimension of the crystal without breaking the periodicity is ~15 mm. The photonic crystal was placed in a rectangular waveguide section 23 × 10 mm. In the course of numerical modelling using CAD ANSYS HFSS, the influence of the change in the parameters of the structure of the photonic crystals

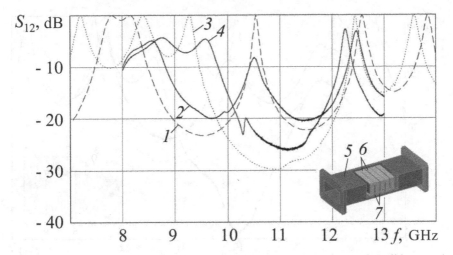

Fig. 1.24. Calculated (dashed and dashed curves) and experimental (solid curves) AFCs of transmittance coefficients S_{12} of eleven-layer low-dimensional PC: curves *1, 2* – with 'broken' sixth central layer of fluoroplastic 1 mm thick; curves *3* and *4* – without breaking the periodicity of the structure. On the inset: a photonic crystal model: a *5*–piece rectangular waveguide, *6*–layers of a dielectric, *7*–capacitive diaphragms [23].

under study on their frequency response was studied in perturbation of the periodicity. Figure 1.24 shows the results of calculation and measurement of the transmission coefficient S_{12} of an eleven-layer photonic crystal without breaking the periodicity (curves *3*, *4*) and with a perturbation of periodicity (curves *1*, *2*) in the form of a central layer of a dielectric (fluoroplastic, $\varepsilon = 2.1$) 1 mm, with a fixed value of the capacitive gap equal to 1 mm.

The measurements were carried out using the Agilent PNA-L Network Analyzer N5230A vector analyzer. Comparison of the results of calculations and measurements indicates their qualitative agreement. The existing difference can be due to the fact that the damping in the walls of the waveguide is ignored, which for the higher types of waves is substantially.

As follows from the results presented in Fig. 1.24, the introduction of a perturbation in the form of a changed size of the central dielectric layer led to the appearance of a defective mode and a significant change in the width and depth of the 'forbidden' band.

Figure 1.25 and 1.26 show the results of calculating and measuring the amplitude–frequency characteristic of the transmission coefficient S_{12} of an eleven-layer photonic crystal with a violation of periodicity for different values of the thickness of the central dielectric layer

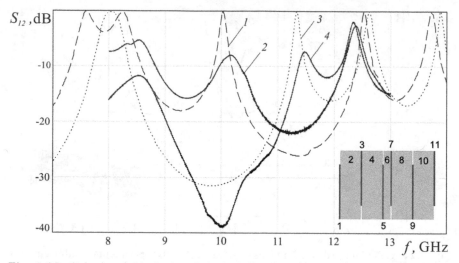

Fig. 1.25. Calculated (dashed and dashed lines) and experimental (solid lines) AFCs of transmittance coefficients S_{12} of the PC with a perturbation in the form of a central layer of a dielectric of different thickness, curves *1*, *2* – thickness of the central layer (layer 6) is equal to 1.5 mm; curves *3*, *4* – thickness of the central layer (layer 6) is 0.5 mm. On the inset: a diagram of a photonic crystal with a 'perturbed' central layer with layer numbers [23].

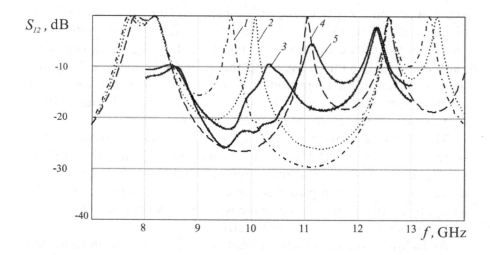

Fig. 1.26. Calculated (dashed, dashed and dot-dash lines) and experimental (solid lines) AFCs of transmittance coefficient S_{12} of a PC with a perturbation in the form of a central layer 1 mm thick. Curve *1* – the magnitude of the capacitive gaps adjacent to the perturbed layer of the left and right metal diaphragms (layers 5, 7 on the inset to Fig. 1.25) is 0.5 mm; curves *2*, *3* – the value of the capacitive gap of the right diaphragm (layer 7 in the inset to Fig. 1.25) is 0.5 mm; curves *4*, *5* – the value of the capacitive gap of the left diaphragm (layer 5 in the inset to Fig. 1.25) is 2 mm [23].

(fluoroplastic) and the magnitude of the capacitive gap of the metal diaphragms adjacent to the 'broken' layer.

From the results shown in Figs. 1.25 and 1.26 it follows that the position of the defect mode on the frequency scale depends significantly not only on the thickness of the 'perturbed' dielectric layer, but also on the capacitance gap of the diaphragms.

Thus, it is shown that a defect mode is possible in a low-dimensional waveguide microwave photonic crystal. High sensitivity to the change in the parameters of the photonic crystal can be used to create small-sized microwave devices with the electrically controlled characteristics and devices for measuring the parameters of materials.

Periodic structures based on resonators as retarding systems for vacuum microwave devices and microwave filters have been described as far back as the 1960s [23, 24]. They were intended to be used as retarding systems in these devices, ensuring the optimal interaction of the electron beam with the electromagnetic wave.

The investigated structure consists of periodically located metal resonant diaphragms at a distance *l* from each other, deposited on a dielectric substrate (Fig. 1.27) [25].

On the basis of numerical simulation using the finite element method in the ANSYS HFSS program, the influence of substrates with different dielectric permittivity on the reflection and transmission coefficients of the microwave wave for the structure was investigated.

From the obtained results it follows that the amplitude–frequency characteristic of the transmittance coefficient of the structure under

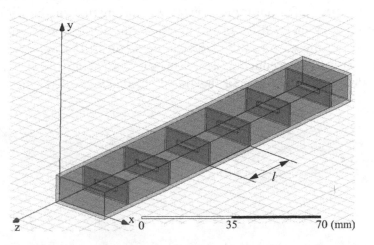

Fig. 1.27. The scheme of a microwave photonic crystal, where *l* is the distance between the diaphragms [26].

study has a 'band' character. The amplitude–frequency characteristics of such a photonic crystal consist of characteristic alternating 'allowed' and 'forbidden' bands. Moreover, the resonances of individual diaphragms can be outside the investigated frequency range, since resonant transmission peaks in a photonic crystal are formed due to the formation of standing waves in the intervals between the diaphragms, and the change in the parameters of the diaphragms (width and height of the slit) allows shifting the 'allowed' and 'forbidden' bands in the desired frequency range.

The AFC of a photonic crystal made up of metal diaphragms deposited on a dielectric substrate with a through slit (Fig. 1.28 *a*) is analyzed for different widths of the aperture and the frequency response of a photonic crystal from diaphragms on dielectric substrates with slits filled with a material with a permittivity ε_2 (see Fig. 1.28 *b*).

It can be noted that the width and depth of the forbidden band increases with increasing gap width *a*, for a fixed dielectric constant of the substrate (Fig. 1.29 *a*). Moreover, the low-frequency edge of the band remains stationary around 9 GHz, and the expansion occurs due to the shift of the high-frequency edge of the forbidden band to the high-frequency region. The same tendency is observed with increasing dielectric permittivity of the dielectric inside the gap (see Fig. 1.29 *b*).

The frequency dependences of the transmission coefficients *D* shown in Fig. 1.29 were obtained by numerical simulation.

The similarity in the behaviour of the amplitude–frequency characteristics of a photonic crystal with a change in the dielectric inside the gap and with a change in the width of the gap is due to the fact that the increase in the dielectric constant of the material inside the gap at a fixed frequency decreases the wavelength in it, which is equivalent to an increase in the linear dimensions of the gap.

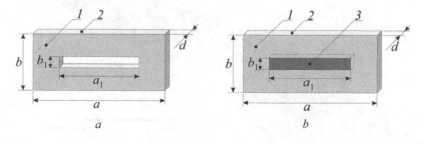

Fig. 1.28. Metal diaphragm *1* on substrate *2* of dielectric with ε_1: *a* – with a through slot; *b* – with a gap filled with a dielectric *3* with a dielectric permittivity ε_2 [26].

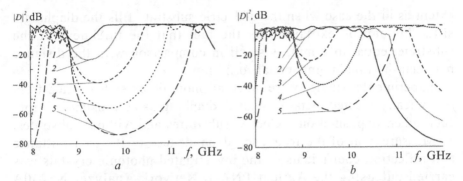

Fig. 1.29. Frequency dependences of $|D|^2$ for a photonic crystal from diaphragms on substrates with $\varepsilon_1 = 4.15$, slit height $b_1 = 2$ mm, dielectric thickness $d = 1$ mm: a – with a through slot of different widths a_1, mm: $1 – 10$; $2 – 11$; $3 – 12$; $4 – 13$; $5 – 14$; b – with a gap of width $a_1 = 10$ mm filled with a dielectric with different dielectric permittivity ε_2: $1 – \varepsilon_2 = 1$; $2 – \varepsilon_2 = 2$; $3 – \varepsilon_2 = 3$; $4 – \varepsilon_2 = 4$; $5 – \varepsilon_2 = 5$ [26].

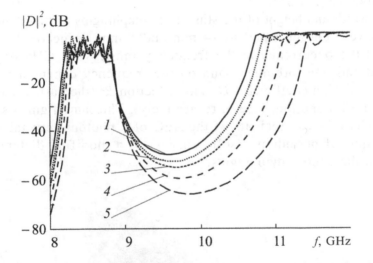

Fig. 1.30. Frequency dependences of the transmission coefficient for a photonic crystal from diaphragms on substrates with a through slot, slit width $a_1 = 10$ mm, slit height $b_1 = 2$ mm, dielectric thickness $d = 1$ mm: 1 – at $\varepsilon_1 = 1$; $2 – \varepsilon_1 = 3$; $3 – \varepsilon_1 = 5$; $4 – \varepsilon_1 = 7$; $5 – \varepsilon_1 = 9$ [26].

The effect of a change in the permittivity of the substrate ε_1 for a photonic crystal from diaphragms with a through slit was studied (see Fig. 1.28, a). Figure 1.30 shows the frequency response of a photonic crystal made up of diaphragms on a dielectric with a through slot, for different values of the permittivity of the substrate. From the results presented in Fig. 1.30, it follows that a change in the permittivity of substrates with through slits does not change the amplitude–frequency characteristic of a photonic crystal to the same extent as in the case when the dielectric substrate fills the diaphragm slit. This can be explained by the fact that the thickness of the substrate, equal to 1 mm, is small in comparison with the distance between the diaphragms equal to 25 mm.

Experimental studies were made of photonic crystals created from periodically located metal resonance diaphragms at a distance *l* from each other, deposited on dielectric substrates and without substrates.

Measurement of the frequency dependences of the transmittance and reflection coefficients of the investigated photonic crystals was carried out using the Agilent PNA-L Network Analyzer N5230A vector analyzer in the frequency range 8–12 GHz.

A photonic crystal based on diaphragms without dielectric substrates consisted of six aluminium diaphragms 10 μm thick fixed between two layers of 2 mm thick polystyrene placed in a rectangular waveguide (Fig. 1.31).

The width and height of the slits of the diaphragms of the photonic crystal were chosen equal to 14 mm and 1 mm, respectively, this ensured the occurrence in the frequency range 8–12 GHz of one allowed and one forbidden band on the frequency dependences of the transmission coefficients D^2 and reflection R^2 (the dashed curves in Fig. 1.32 *a* and Fig. 1.32 *b*, respectively). The same figures show the results of measurements of the AFC of a photonic crystal with a violation of periodicity (solid curves) as a modified distance L between the central diaphragms.

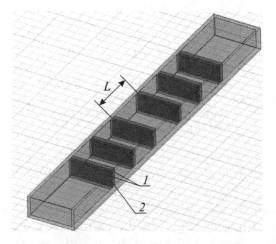

Fig. 1.31. Diagram of a photonic crystal based on diaphragms without dielectric substrates: *1* – polystyrene, *2* – metal diaphragm, L – size of disturbance in the central layer [26].

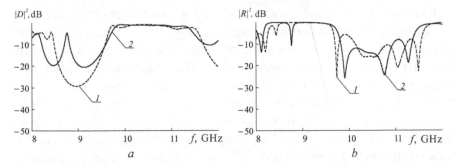

Fig. 1.32. Experimental amplitude–frequency characteristics of the transmission coefficient (*a*) and reflection coefficient (*b*) of a photonic crystal based on diaphragms without dielectric substrates and without perturbation (curve *1*) and with a perturbation of periodicity (curve *2*). The distance between the diaphragms is $l = 27$ mm, the slit length is $a_1 = 14$ mm, the perturbation size in the central layer is $L = 20$ mm [26].

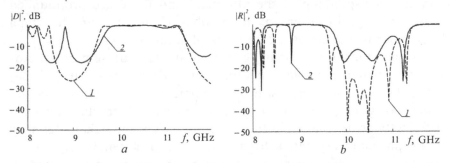

Fig. 1.33. Calculated amplitude–frequency characteristics of the transmission coefficient (*a*) and reflection coefficient (*b*) of a photonic crystal based on diaphragms without dielectric substrates without perturbation (curve *1*) and with a perturbation of periodicity (curve *2*). The distance between the diaphragms is $l = 27$ mm, the slit length is $a_1 = 14$ mm, the perturbation size in the central layer is $L = 20$ mm [26].

The results of calculating the AFC of a photonic crystal based on diaphragms without dielectric substrates with violation and without violation of periodicity with parameters corresponding to the experimental sample described above are shown in Fig. 1.33 *a, b*.

Comparison of the experimental dependences presented in Fig. 1.32 with the results of the calculation shown in Fig. 1.33 indicates a good qualitative and quantitative correspondence. In this case, as follows from the experiment and the results of the calculations, the creation of a perturbation in a photonic crystal in the form of a reduced distance between the central diaphragms leads to the appearance of a transmission peak in the forbidden band and an increase in its width.

Experimental studies were made of photonic crystals created from periodically located at a fixed distance from each other metal resonance diaphragms deposited on solid dielectric substrates and dielectric substrates with through slits. The experimental frequency response of a photonic crystal created from periodically located at a fixed distance from each other metal resonant diaphragms deposited on solid dielectric substrates is shown in Fig. 1.34 *a*. The results of the calculation of the AFC of a photonic crystal based on periodically located diaphragms with solid dielectric substrates with parameters corresponding to the experimental sample described above are shown in Fig. 1.34 *b*.

The results presented in Figs. 1.34 *a* and *b* show their good qualitative and quantitative correspondence, while in the frequency domain corresponding to the allowed band there are pronounced transmission peaks observed, the number of which is one less than the number of diaphragms forming a photonic crystal .

The methods for measuring the parameters of semiconductor structures using microwave photonic crystals described in [26–28] assumed that the measured structure completely fills the cross section of the waveguide. Since in this case the measured structure causes a sufficiently sharp change in the amplitude–frequency characteristics of the photonic crystal, a high resolution and sensitivity to the change in the parameters of the measured structures is provided. However, the locality of the measurements is thus limited to the cross-sectional area of the waveguide, and to increase the locality of the measurements, it is required to use waveguides of a higher frequency range, which, as a rule, causes certain technical difficulties.

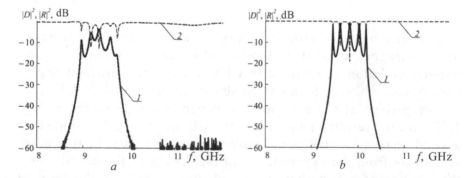

Fig. 1.34. AFC of a photonic crystal based on periodically located diaphragms with solid dielectric substrates of the reflection (curve *2*) and transmission coefficients (curve *1*). The distance between the plates is $l = 25$ mm, the slit width is $a_1 = 14$ mm, the slit height is $b_1 = 2$ mm: *a* – is the experiment; *b* – calculation [26].

In the case when the investigated structure is only partly filling the waveguide cross section, it is of interest to study the influence of its size and location within the previously created inhomogeneity, which acts as a microresonator in a photonic crystal to enhance the response of the amplitude–frequency characteristics of a photonic crystal [29]. The authors of [11] showed that the creation of a large number of small air inclusions inside the disturbed layer leads to a change in the frequency response of the transmission coefficient of a photonic crystal, expressed in particular in the shift of the transmission peak in the forbidden band toward higher frequencies, that is, in the change in the frequency of the defect mode. In this case, it is possible to describe the AFC of a photonic crystal using the model of an 'effective' medium. As an alternative to air inclusions for controlling the AFC of a photonic crystal, it is possible to consider the introduction into the perturbed layer of metallic inclusions and the associated change in the characteristics of the defect mode.

The authors of Ref. [30] have studied the frequency response of the transmission coefficient of a microwave photonic crystal with a small-sized, planar conducting inclusion introduced into the perturbed layer and partially filling the cross section of a waveguide.

A waveguide photonic crystal consisting of eleven layers was studied in the frequency range 8–12 GHz (Fig. 1.35). Odd layers were made of ceramics (Al_2O_3, $\varepsilon = 9.6$), even layers were made of fluoroplastic ($\varepsilon = 2.0$). The thickness of odd layers $d_{Al_2O_3} = 1$ mm, even layers $d_{tef} = 22$ mm. The layers completely filled the cross section of the waveguide.

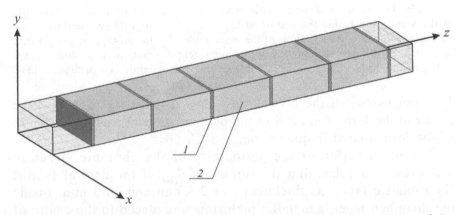

Fig. 1.35. 3D model of a waveguide photonic crystal without perturbation of the periodicity: *1* – ceramic (Al_2O_3), *2* – fluoroplastic [31].

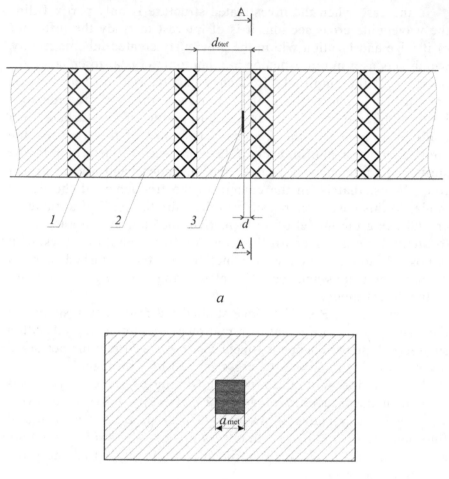

Fig. 1.36. The position of the metallic inclusion in the form of an aluminium film in the waveguide inside the disturbed layer: *a* – the longitudinal section of the waveguide, *b* – the cross section of the waveguide: *1* – the Al_2O_3 ceramic layer, 1 mm thick; *2* – a layer of fluoroplastic, 22 mm thick; *3* – metallic inclusion, located in the perturbed layer, remote from the layer of Al_2O_3 ceramic at a distance *d* [31].

The dimensions of the layers were chosen in such a way that the center of the forbidden band of the photonic crystal was in the centre of the investigated frequency range 8–12 GHz.

The perturbation of the periodicity of the photonic structure was created by changing the thickness d_{6tef} of the central (sixth) fluoroplastic layer, its thickness was 2.3 mm and 14.5 mm. Inside the disturbed layer, a metallic inclusion was placed in the centre of the waveguide section in the form of an aluminium film 100 μm thick, having the shape of a square with the side a_{met} equal to 3 mm (Fig. 1.36).

On the basis of numerical simulation using the software for three-dimensional modelling of electromagnetic fields by the ANSYS HFSS finite element method, the amplitude-frequency characteristics of the transmission coefficient of a photonic crystal with a perturbation of periodicity were investigated in the presence of a metallic inclusion introduced inside the disturbed layer.

As noted above, the presence of a perturbation of the periodicity of the photonic crystal leads to the appearance of a defect mode on the frequency response of the transmittance coefficient.

The distribution of the electric field strength of the electromagnetic wave at the frequency of the defect mode was calculated to study the effect of the location of a metallic inclusion in the form of an aluminium film inside a disrupted layer that acts as a microresonator in a photonic crystal, on the characteristics of the defect mode or the 'transparency window'.

The results of calculating the electric field strength of an electromagnetic wave show the alternation of nodes and antinodes inside a photonic crystal along the direction of its propagation. In this case, for the selected parameters of the photonic crystal without metallic inclusion at the frequency of the defect mode f_{res}, a node is observed in the centre of the perturbed layer (curve 5 in Fig. 1.37).

As follows from the dependences presented in Fig. 1.37, the introduction of a metallic inclusion in the form of an aluminium film inside the broken layer of a photonic crystal leads to distortion in the distribution of the electric field strength, while the maximum perturbation is observed when the metallic inclusion is located on the boundary of the disturbed layer.

Such a change in the distribution of the electric field strength can lead to a change in the position of the resonant feature in the forbidden band of a photonic crystal.

The results of the calculation of the AFC of a photonic crystal at selected positions of a metallic inclusion with a size of $a_{met} = 3$ mm inside the perturbation are shown in Fig. 1.38. The inset in Fig. 1.38 shows the frequency response of a photonic crystal near the frequency of the defect mode in the forbidden band of a photonic crystal.

The analysis presented in Fig. 1.38 dependences shows that the presence of a metallic inclusion in the inhomogeneity leads to a shift in the frequency of the defect mode f_{res} of the photonic crystal to the

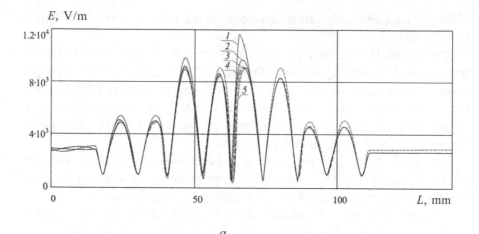

Fig. 1.37. The distribution of the electric field strength of an electromagnetic wave inside a photonic crystal for a different position of the metallic inclusion d inside the inhomogeneity and without inclusion: a – along the direction of propagation of the electromagnetic wave; b – in the transverse plane of the waveguide (A–A) in Fig. 1.3 b, a. f_{res} = 10.095 GHz. The thickness of the perturbed layer is d_{6tef} = 2.3 mm. d, μm: 1 – 0, 2 – 100, 3 – 200, 4 – 300, 5 – structure without metallic inclusion [31].

low-frequency region relative to the frequency of the defect mode f_0 in the absence of inclusion in the inhomogeneity. In this case, the magnitude of the shift $\Delta f = f_{res} - f_0$ is maximal (Δf = 63 MHz) when the inclusion is located at the boundary of the disturbed layer $d = 0$ and decreases as it shifts to the center of the inhomogeneity (Δf = 13 MHz at d = 300 μm), Δf is minimal when the metallic inclusion is located at the centre of the inhomogeneity d = 1.15 mm.

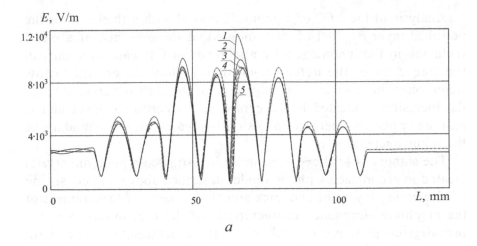

a

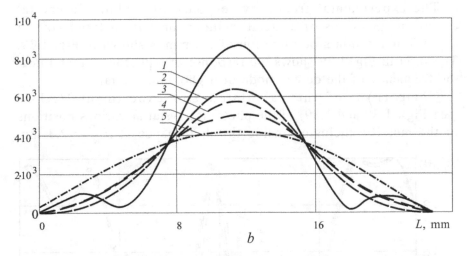

b

Fig. 1.38. AFC of a photonic crystal at various positions *d* within the violation of a metallic inclusion with a size of a_{met} = 3 mm. *d*, μm: *1* – 0, *2* –100, *3* – 200, *4* – 300, *5* – structure without metallic inclusion [31].

Calculations show that the defect mode created with a thickness d_{6tef}=2.3 mm (d_{6tef}< λ_{res}/2, where λ_{res} is the wavelength at the frequency of the defect mode) can be realized even at a larger thickness d_{6tef}=14.5 mm (λ_{res}/2< d_{6tef}<λ_{res}), however, in this case there is a significant change in the electric field intensity distribution of the electromagnetic wave inside the photonic crystal along the direction of its propagation at the frequency of the defect mode, consisting in the emergence of an antinode of the electric field strength in the center of the perturbed layer.

Analysis of the AFC of a photonic crystal with a thickness of the pertubed layer d_{6tef} = 14.5 mm shows that the presence of a metal inclusion in the inhomogeneity also leads in this case to a shift in the frequency of the defect mode of the photonic crystal to low frequencies. In this case, the magnitude of the shift is maximal when the inclusion is located in the centre of the perturbed layer and is minimal when the metallic inclusion is located on the boundary of the inhomogeneity.

The authors of [30] experimentally investigated a photonic crystal created in accordance with the model described above and consisting of alternating layers of ceramics and fluoroplastic. Measurement of the amplitude–frequency characteristic of the transmittance of the investigated photonic crystal in the three-centimeter wavelength range was carried out using the Agilent PNA-L Network Analyzer N5230A vector analyzer.

The experimental frequency response of a photonic crystal at various positions of a metallic inclusion with a fixed size of a_{met} = 3 mm within a perturbation of 2.3 mm is shown in Fig. 1.39. The inset in Fig. 1.39 shows the response of a photonic crystal near the frequency of the defect mode of a photonic crystal.

Comparison of the calculated and experimental AFCs (see Figs. 1.38 and 1.39) of the photonic crystal at various positions of the metallic inclusion inside the disturbance at d_{6tef} = 2.3 mm

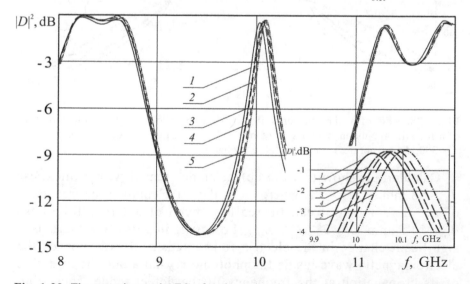

Fig. 1.39. The experimental AFC of a photonic crystal at various positions d within the violation of a metallic inclusion with a size of a_{met} = 3 mm. d, μm: 1 – 0, 2 – 100, 3 – 200, 4 – 300, 5 – structure without metallic inclusion [31].

indicates their good quantitative coincidence and confirms that the magnitude of the shift is maximal when the inclusion is located in the centre of the perturbed layer and is minimal when the metallic inclusion is located on the boundary of the inhomogeneity.

The AFC of a photonic crystal with an increased thickness of the central perturbed layer equal to 14.5 mm was measured and the analysis results confirmed the computer simulation data, which indicates that the magnitude of the defect mode shift in this case is maximal when the metallic inclusion is located at the inhomogeneity boundary and is minimal when the metallic inclusion is located in the centre of the perturbed layer .

From the results obtained by the authors of [30] it follows that the introduction of a small planar conducting inclusion into the broken layer of a perturbed crystal leads to a shift of the defect mode in the forbidden band toward lower frequencies, while the maximum shift is observed when the metallic inclusion is located at the boundary of the perturbation at a thickness of the perturbed layer $d_{6tef} < \lambda_{res}/2$ and in the centre of the perturbation at $\lambda_{res}/2 < d_{6tef} < \lambda_{res}$. The obtained results can be used, in particular, for the design of microwave photonic crystals with electrically controlled characteristics.

References

1. Gomez A., Vegas A., Solano M.A. & Lakhtakia A., Electromagnetics. 2005. Vol. 25, issue 5. P. 437–460.

2. Usanov D.A., Skripal' A.V., Abramov AV, Bogolyubov A.S., Zh. Teor. Fiz., Zh. 2006. V. 76, No. 5. P. 112–117.

3. Chaplygin Yu.A., Usanov D.A., Skripal' A.V., Abramov A.V., Bogolyubov A.S., Izv. VUZ, Elektronika. 2006, No. 6, P. 27–35.

4. Usanov D.A. et al., Proc. of 36rd European Microwave Conference. Manchester, UK. 10–15th September 2006. 509–512.

5. Usanov D.A., et al., Radiotekhnika i elektronika. 2013. Vol. 58. No. 11. P. 1071–1076.

6. Yablonovitch E., Gimitter T. J., Meade R.D., Phys. Rev. Lett. 1991. Vol. 67, No. 24. P. 3380–3383.

7. Belyaev B.A., Voloshin A.S., Shabanov V.F., Dokl. Akad. Naok, 2005. V. 403, No. 3, P. 319–324.

8. Gunyakov V.A., et al., Pis'ma Zh. Teor. Fiz. 2006. Vol. 32, No. 21. P. 76–83.

9. Usanov D.A., et al., Izv.VUZ, Elektronika. 2010. No. 1. P. 24–29.

10. Gunyakov V.A., Zh. Teor. Fiz. 2010. V. 80, No.. 10. P. 95–100.

11. Usanov D.A., et al., ibid, 2016. Vol. 86, No. 2. 65–70.

12. Maxwell-Garnett J. C., Philos. ransactions of the Royal Society, 1904, V. 203, P. 385–420.

13. Bruggeman D.A., Annalen der Physik, 1935, V. 24, No. 8, P. 636–679.

14. Lichtenecker K., Physikalische Zeitschrift, 1926, V. 27, No.4, P. 115–158.

15. The patent of the Russian Federation for the invention № 2360336. Wide-band waveguide consistent load. Usanov D.A., et al., Ppubl. 27.06.2009. Bul. No. 18.

16. Usanov D.A.,et al., Izv. VUZ, Elektronika, 2009. No. 1. P.73 – 80.

17. Katsumi Yoshino, et al., IEEE Transactions on Dielectrics and Electrical Insulation. June 2006. Vol. 13, No. 3. P. 678–686.

18. Belyaev B.A., Voloshin A.S., Shabanov V.F., Dokl. Akad. Nauk. 2005. V. 400, No. 2. P. 181–185.

19. Fernandes H.C.C., Medeiros J.L.G., Junior I.M.A., and Brito D.B. Photonic Crystal at Millimeter Waves Applications//PIERS Online. 2007. Vol. 3, no. 5. P. 689–694.

20. Ozbay E., Temelkuran B., Bayindir M., Progress in Electromagnetics Research, 2003. Vol. 41, P. 185–209.

21. Saib A., Huynen I., Electromagnetics. 2006. Vol. 26, issue 3–4, p. 261–277.

22. Gulyaev Yu.V., et al., Dokl. Akad. Nauk, 2014. V. 458, No. 4. P. 406–409.

23. Silin R.A., Sazonov V.P., Slowing systems. Moscow: Sov. radio, 1966, 631p.

24. Cohn S.B., Proc. IRE. Feb 1957. Vol. 45. P. 187–196.

25. Usanov D.A., Radiotekhnika, 2015. No. 10. P. 108–114.

26. Usanov D.A.,et al., Izv. VUZ, Elektronia. 2007. No. 6. P. 25–32.

27. Gulyaev Yu.V., et al., Dokl. Akad. Nauk, April 2012. V. 443, No. 5. P. 564–566.

28. Usanov D.A., et al., Proceedings of the 44th European Microwave Conference. Rome, Italy. 6-9 Oct 2014. P. 984–987.

29. Joannopoulos I.D., Villenneuve Pierre R., Fan Shanhui, Nature. 1997. Vol. 386, No. 13. P. 143–149.

30. Usanov D.A., Skripal' A.V., Romanov A.A., Zh. Tekh. Fiz., 2017. V. 87, No. 6. P. 884–887.

2

Microwave photonic crystals based on planar transmission lines

The properties of one-dimensional periodic structures with a photonic band gap on coplanar waveguides, coplanar strip lines and slot lines have been investigated by the authors [1, 2]. The periodicity in the change in the waveguide resistance was provided by a corresponding change in the size of the gaps. The possibility of creating a band-stop microwave filter and a microwave resonator of the Fabry–Perot type with reflectors on both sides was demonstrated. Images of the investigated structures with photon bands are shown in Fig. 2.1 *a*. Figure 2.1 *b* shows the images of the resonators on these lines.

The created resonators are based on the use of defects introduced into periodic structures. The length of the structure element *d* was determined by the relation $k = \dfrac{\pi}{d} = \dfrac{2\pi}{\lambda}$, where *k* is the wave propagation constant in the line. Hence, *d* is equal to half the wavelength in the waveguide. As the material of the substrates of plane lines, the authors of Ref. 1 used a dielectric with $\varepsilon_r = 10.5$. The number of cells was chosen equal to seven. The results of numerical simulation and measurements for photonic structures and resonators, given in [1], are shown in Fig. 2.2.

The width of the 2.8 GHz stop band was reached with the reflection in this band of −36.5 dB (Fig. 2.2 *a*). For an unloaded resonator, a *Q*-factor of 299.1 was obtained (see Fig. 2.3 *c*).

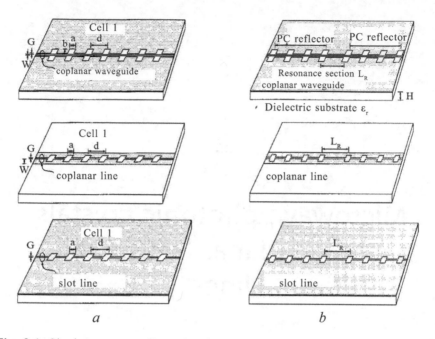

Fig. 2.1. Single-plane one-dimensional periodic structures (coplanar waveguide, coplanar strip line, slot line) with a photonic band gap: *a* – one-dimensional blocking filters, *b* – one-dimensional resonators [1].

The characteristics of a structure with a photonic band gap on the basis of a coplanar line were investigated by the authors of [2]. As a substrate material, they used thin ferroelectric films of composition $Ba_{0.8}Sr_{0.2}TiO_3$. The structure of the investigated photonic crystal was a sequence of segments of a coplanar waveguide with a wave resistance of 50 and 20 ohms. The dimensions were determined as a result of preliminary numerical calculations of the reflection and transmission coefficients. Figure 2.4 shows the results obtained by the authors of Ref. [2] for calculating the frequency dependences S_{11} and S_{21} and the design of the line (the dimensions of the structural elements are given in mm). Figure 2.5 shows the results of measurements of the dependence of the reflection and transmission coefficients, consistent with the calculation.

The authors of [3] noted the possibility of using flat microwave structures with a photonic band gap for suppression of higher-order modes in active and passive devices. The construction of the proposed line is shown in Fig. 2.6.

It is shown in [3] that the parameters acceptable for practice can be provided using a line with periodically etched circular holes under

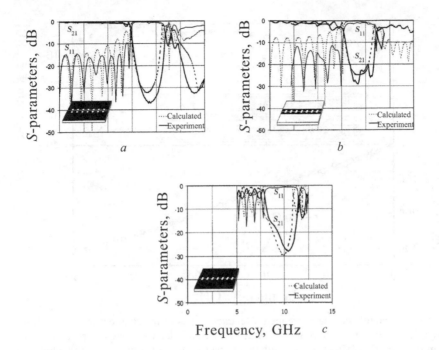

Fig. 2.2. Calculated and measured reflection and transmission spectra for stop band filters based on single-plane one-dimensional periodic structures: *a* – coplanar waveguide, *b* – coplanar strip line, *c* – slot line [1].

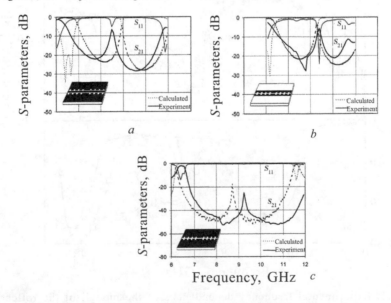

Fig. 2.3. Calculated and measured reflection and transmission spectra for resonators with reflectors based on single-plane one-dimensional periodic structures: *a* – coplanar waveguide, *b* – coplanar strip line, *c* – slot line [1].

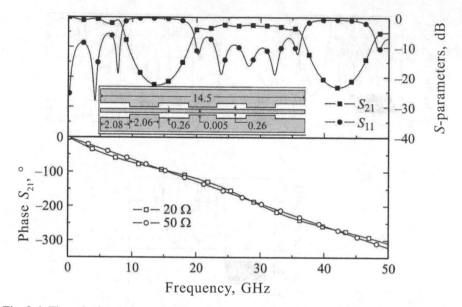

Fig. 2.4. The calculated phase–frequency characteristics of the segments of the coplanar waveguide forming the photonic crystal period and the frequency dependences of the moduli of the reflection S_{11} and transmission S_{21} coefficients of the photonic crystal. On the inset: the topology of a photonic crystal [2].

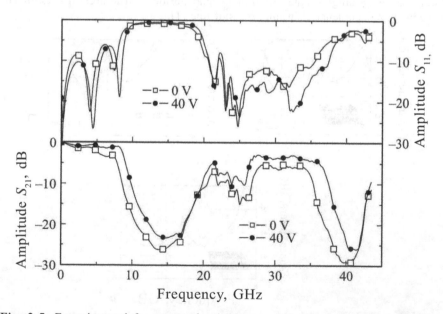

Fig. 2.5. Experimental frequency dependences of the moduli of the reflection coefficients S_{11} and transmission coefficients S_{21} of the photonic crystal obtained in the absence of bias voltage (0 V) on the $Ba_{0.8}Sr_{0.2}TiO_3$ film and in the case of a bias voltage of 40 V applied to the film [2].

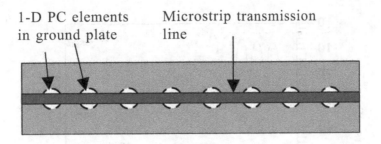

1-D PC elements   Microstrip transmission
in ground plate   line

Fig. 2.6. Geometric configuration of a one-dimensional structure with a photonic band gap with a uniform distribution of cylindrical holes located under the 50-ohm transmission line [3].

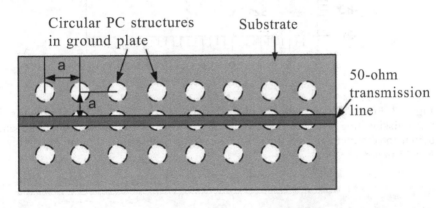

Circular PC structures   Substrate
in ground plate

50-ohm
transmission
line

Fig. 2.7. Geometric configuration of a standard 50-ohm transmission line located above three lines of holes in a two-dimensional structure with a photonic band gap [3].

the strip. The design with 1-D located holes provides parameters close to a similar design with a 2-D system of holes (Fig. 2.7), as follows from the measurement results shown in Fig. 2.8.

The authors of [4] demonstrated the possibility of obtaining a photonic stop band on a microwave using a microstrip line with periodically located rectangular apertures (Fig. 2.9). For the strip line, the authors of [4] used an aluminium foil 18 µm thick. The distance between the metal strips and the base plane was ~1 mm.

The authors of [5] proposed a calculation method and measured the amplitude–frequency characteristics of the Bragg structure in the form of a microstrip line with a different width of strip conductors. In the frequency band 1–2.5 GHz, the bandwidth of $\frac{\Delta f}{f_0}$ 38.8% was

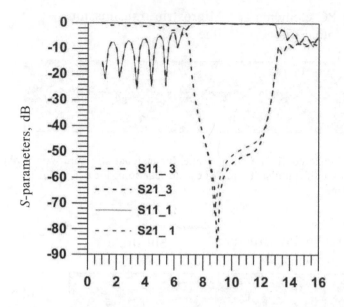

Fig. 2.8. Frequency dependences of S-parameters of a standard 50-ohm microstrip transmission line with a ground plate in the form of two-dimensional and one-dimensional structures with a photonic band gap. Substrate – Taconic, $\varepsilon_r = 10$ and $h = 25$ mm [3].

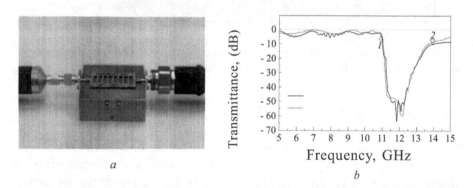

a   *b*

Fig. 2.9. Photo of the experimental structure (*a*). The period of the array of rectangular holes is 4 mm, the dimensions of the rectangular hole are 3 mm×12 mm, the width of the metal strip is 20 mm; *b* – measured and calculated transmission spectra for the case of filling the substrate with air. To measure the transmission spectra, the network analyzer HP 8753 Network Analyzer was used, the method of moments was used for calculations. Two markers show approximate limits of the band gap [4].

reached, where $f_0 = \sqrt{f_1 f_2}$ is the central frequency of the passband. $\Delta f = f_2 - f_1$.

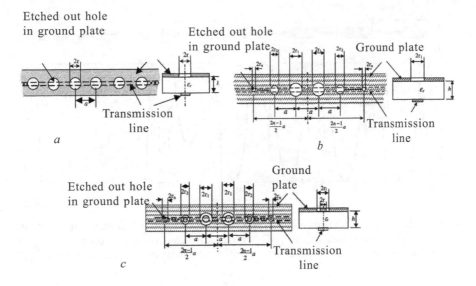

Fig. 2.10. Structure with a photonic band gap (*a*) with a uniform distribution of cylindrical holes with period *a* and radius *r*; *b* – structures with a photonic band gap with a non-uniform distribution of cylindrical and (*c*) non-uniform distribution of annular holes. Number of holes – even [6].

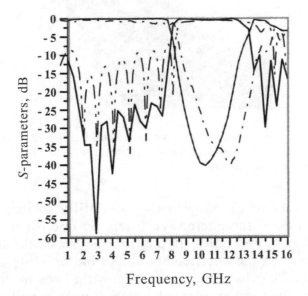

Fig. 2.11. Frequency dependences of *S*-parameters for a structure with a photonic band gap with a uniform linear distribution of 10 annular holes (type 1) with a Chebyshev polynomial side-lobe distribution of 25 dB with *r* = 84 mm, period *a* = 224 mm and line width 24 mm (50 ohms). Dielectric substrate: ε_r = 10.2, *h* = 25 mm. Solid line – calculation, double dash-dot line – measurement results for side lobe level at −25 dB [6].

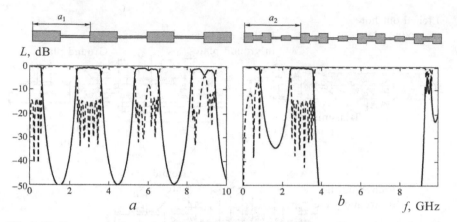

Fig. 2.12. Frequency dependences of microstrip models of one-dimensional photonic crystals: *a* – with a standard lattice; *b* – with a lattice consisting of two sublattices. The solid and dashed lines correspond to the introduced and reverse losses, respectively [7].

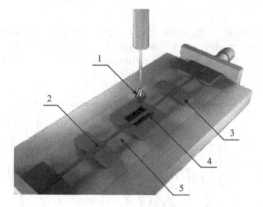

Fig. 2.13. Microstrip photonic crystal with a cuvette for measuring the parameters of liquid dielectrics: *1* – measured liquid, *2* – strip conductor, *3* – Al_2O_3 ceramics, *4* – measuring cell, *5* – air gap [8].

The authors of [6] showed the possibility of improving the characteristics of the microwave filters on microstrip lines with periodically located holes, if the dimensions of the holes vary in proportion to the coefficients of the binomial or Chebyshev polynomials. The characteristics of microstrip lines for ten-element filters with round and annular holes etched in the metallized base of the line were investigated. Figure 2.10 shows the various designs. Filters with a uniform distribution of holes have the same radii and the distance between the holes. For an inhomogeneous distribution, the central element has a larger dimension r_1, which is proportional to the coefficient 1, the radii of neighbouring circles have discretely proportional amplitude coefficients of the polynomials. The polynomial coefficients were chosen proportional to the radius of the circle (type 1) or the area of the circle (type 2). In the study, the authors of [6] used a dielectric with $\varepsilon_r = 10.2$. The central frequency was chosen to be 10 GHz.

When using the binomial distribution, as the calculation shows, a relatively narrow stop band and low-selectivity is obtained. The characteristics of the filter can be significantly improved by using the Chebyshev distribution of inhomogeneities, as follows from the results of the simulation and measurements shown in Fig. 2.11. The angularity of the frequency characteristics in the passband has a reduced value, while for a uniform distribution, the angularity in the bandwidth region is very high. However, the width of the locking band is somewhat lower.

The authors of Ref. [7] investigated the quality factor of the resonance of the impurity mode in a one-dimensional microstrip photonic crystal. The model of a photonic crystal of this type is based on a strong dependence of the effective dielectric constant of the microstrip transmission line on the width of its strip conductor and the thickness of the substrate. The investigated photonic crystal consisted of successively connected alternating line segments with a 'large' and 'small' width of the strip conductor. The topology of the conductors of the microstrip crystal investigated by the authors is shown in Fig. 2.12 with the usual lattice and two sublattices. The same figure shows the amplitude–frequency characteristics given in [7]. The authors of [7] note that the stop band in a crystal with two sublattices can be several times wider than in a crystal with an ordinary lattice.

The characteristics of microstrip photonic crystals were calculated and experimentally investigated by the authors [8]. A structure was studied that is a series of connected segments of a microstrip transmission line with periodically varying dielectric permittivity of the substrate. Even segments are realized on a Al_2O_3 ceramic substrate (Al_2O_3), and odd segments are in the form of strip line segments with air filling, in which an air gap is created between the strip and the metal base. The image of the microstrip photonic structure under consideration is shown in Fig. 2.13.

To calculate the coefficient of transmission and reflection of an electromagnetic wave, a transmission matrix **T** of a four-terminal network of complex structure was used in the quasistatic approximation, which is a cascade connection of elementary four-ports with known transmission matrices, which have the form:

$$\mathbf{T} = \begin{pmatrix} T[1,1] & T[1,2] \\ T[2,1] & T[2,2] \end{pmatrix} = \mathbf{T}_N' \cdot \overset{N-1}{\underset{i=1}{Đ}} \left(\mathbf{T}_{i,i+1}'' \cdot \mathbf{T}_i' \right), \qquad (2.1)$$

where $\mathbf{T}_i'$ and $\mathbf{T}_{i,j}''$ are the transmission matrices of the four-ports, respectively, the i-th segment and the direct connection of the i-th and ($i+1$)-th segments of the microstrip transmission line.

The expressions for the transmission matrices $\mathbf{T}_i'$ and $\mathbf{T}_{i,j}''$ of the corresponding elementary four-ports are:

$$\mathbf{T}_i' = \begin{pmatrix} e^{\gamma_i d_i} & 0 \\ 0 & e^{-\gamma_i d_i} \end{pmatrix}, \qquad (2.2)$$

$$\mathbf{T}_{i,i+1}'' = \begin{pmatrix} \dfrac{r_{i,i+1}+1}{2\sqrt{r_{i,i+1}}} & \dfrac{r_{i,i+1}-1}{2\sqrt{r_{i,i+1}}} \\ \dfrac{r_{i,i+1}-1}{2\sqrt{r_{i,i+1}}} & \dfrac{r_{i,i+1}+1}{2\sqrt{r_{i,i+1}}} \end{pmatrix}. \qquad (2.3)$$

Here d_i is the length of the i-th segment, γ_i is the propagation constant of the electromagnetic wave in the i-th segment, $r_{i,i+1} = \dfrac{\rho_{i+1}}{\rho_i}$,

where $\rho_i = \dfrac{377 \cdot h_i}{\sqrt{\varepsilon_i} \cdot W_i \cdot \left(1 + 1.735 \cdot \varepsilon_i^{-0.0724} \cdot \left(\dfrac{W_i}{h_i} \right)^{-0.836} \right)}$ — wave resistance of

the i-th length of the structure, ε_i, h_i is the dielectric constant and thickness of the substrate of the i-th segment, W_i is the width of the strip conductor of the i-th segment.

The transmission coefficient of the microwave power is determined through the transmission matrix element $T[1,1]$ T:

$$D = \frac{1}{|T[1,1]|^2}. \qquad (2.4)$$

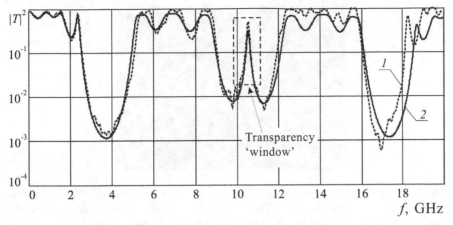

Fig. 2.14. Calculated frequency dependences of the transmittance of the microstrip photonic structure with a violation of periodicity in the form of the changed lengths of the 4th, 5th and 6th segments (curve *1*) and without violation of periodicity (curve *2*) [9].

The reflection coefficient of the microwave power from the microstrip photonic structure is determined by the square of the modulus of the element $S[1,1]$ of the scattering matrix

$$R = |S[1,1]|^2, \qquad (2.5)$$

whose elements are associated with the elements of the transmission matrix **T** by the relation:

$$\mathbf{S} = \begin{pmatrix} S[1,1] & S[1,2] \\ S[2,1] & S[2,2] \end{pmatrix} = \begin{pmatrix} \dfrac{T[2,1]}{T[1,1]} & \dfrac{T[1,1] \cdot T[2,2] - T[1,2] \cdot T[2,1]}{T[1,1]} \\ \dfrac{1}{T[1,1]} & -\dfrac{T[1,2]}{T[1,1]} \end{pmatrix}. \qquad (2.6)$$

The authors of [9] studied a photonic structure consisting of seven successively connected alternating segments of a microstrip transmission line with a large and small width of the upper conductor included in the 50-ohm transmission line. The width of the first, third, fifth and seventh segments of the strip line was 2.5 mm. The second, fourth, sixth segments – 0.5 mm. The length of wide segments was 7 mm, narrow – 7.6 mm.

Figure 2.14 shows the experimental and calculated frequency dependences of the transmittance of a photonic crystal based on a 7-layer microstrip structure [9]. The violation of periodicity was created in the form of a shorter length of the fourth high-resistance segment of the microstrip transmission line (l_4 = 5.1 mm). The

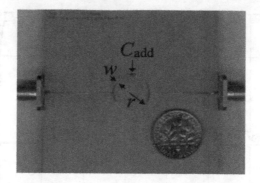

Fig. 2.15. Photo of the created microstrip structure with a photonic band gap with the chip capacitor installed on it. The mean radius of the ring is ε_m, the width of the ring line is w, the gap size is g 6.25 mm, 0.5 mm and 0.2 mm, respectively [12, 13].

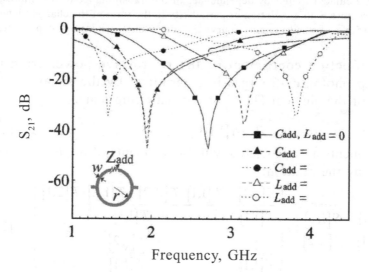

Fig. 2.16. The transmission spectrum for the created microstrip structure with photonic band gap for different values of the installed capacitances and inductances. The inset shows a schematic representation of a configurable microstrip structure with a photonic band gap [12, 13].

frequency dependence of the transmission coefficient of the simulated microstrip photonic structure is characterized by the presence of frequency domains that are forbidden for the propagation of an electromagnetic wave-analogues of forbidden bands in crystals. The presence of a violation of periodicity led to the appearance of a «window» of transparency in the forbidden band of a photonic crystal with a central frequency of 10.5 GHz.

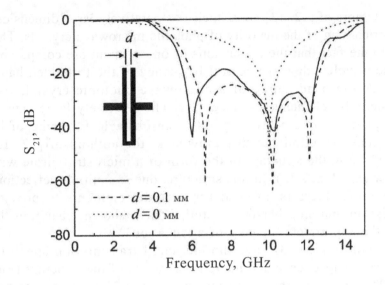

Fig. 2.17. Measured and calculated transmission spectra for the structure represented on the insert with an average circumference of $l = 14$ mm, $w = 0.3$ mm and $d = 0.1$ mm. The structure demonstrates a broad forbidden band with three peaks of attenuation [13].

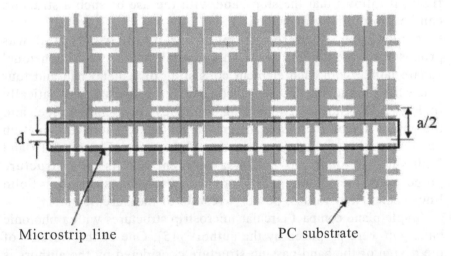

Microstrip line   PC substrate

Fig. 2.18. Schematic representation of a microstrip transmission line with an ground plate in the form of a single-plane compact structure with a photonic band gap [14].

Referring to [10, 11], the authors of [7] expressed the opinion that it is convenient to investigate the possibility of increasing the Q-factor of the resonant impurity mode on irregular microstrip structures.

The authors of [12, 13] noted their comparatively large dimensions as a shortcoming of the majority of photonic microwave crystals. This is due to the fact that the dimensions of one element are comparable with the wavelength at the central frequency of the forbidden band. To realize the band nature of a microwave photonic crystal, five or six periodic elements are necessary. The relatively large size of such a photonic microwave crystal is an obstacle for some of its applications. To eliminate this drawback, the authors of [12, 13] proposed using the structure in the form of a microstrip circle with a narrow gap (Fig. 2.15). In this structure, due to multiple reflections from the gap, there is no need for periodic elements to provide transmission and stop bands. As studies have shown, changing the size of the gap affects the pole of attenuation.

The authors [12, 13] gave the measured transmission spectrum for different capacitance and inductance values of the structure (Fig. 2.16). The centre of the spectrum with f_1 = 2.71 GHz is obtained without additional built-in capacitance and inductance.

The modification of the proposed structure and its spectrum S_{21}, given in [12], are shown in Fig. 2.17. From the results shown in this figure it follows that the stop band with the use of such a structure can be substantially expanded.

A new type of microwave photonic microstrip crystal was proposed by the authors [14]. This crystal was a two-dimensional square lattice with each element consisting of a metal pad and four connecting branches. The proposed structure is shown schematically in Fig. 2.18. Narrow connecting branches form inductance, and the gaps between the platforms are capacitance. Their combination determines the propagation constant of the wave. Figures 2.19 and 2.20 show photographs and measurement results of a structure proposed by the authors of [14] connected to microstrip lines. Solid line – calculated values, points – results of measurements.

Single-plane compact circular microstrip structures with a photonic band gap were proposed by the authors [15]. One of the variants of the design of the band-passing structure considered by the authors is shown in the inset in Fig. 2.21. Photos of the structures are shown in Fig. 2.22. To ensure the minimum dimensions of the elements of the photonic crystal, it uses a combination of curved inner strips with circles etched in the main plane. The results of numerical simulation of the characteristics of this structure and measurements are shown in Fig. 2.21.

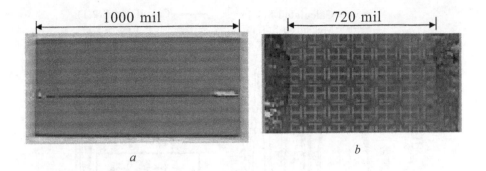

Fig. 2.19. Photos of created microstrip transmission lines with a ground plate in the form of a single-plane compact structure with a photonic bandgap: *a* – top view, *b* – bottom view [14].

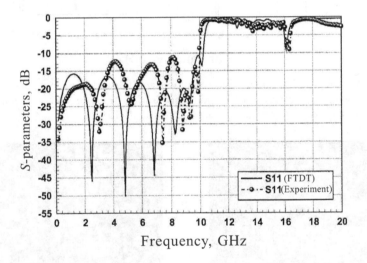

Fig. 2.20. Comparison of the calculated and experimentally obtained reflection coefficients S_{11} [14].

A new type of microwave photonic crystal is described in [16]. In it, a waveguide-slot line is used as the Bragg structure, which has a periodically varying dielectric constant in space. Such a line is interesting in that its application opens the prospect of creating microwave integrated circuits with unique characteristics [17], in this case related to the properties of the Bragg structures [17].

A general view of the Bragg microwave structure based on the waveguide-slot lines is shown in Fig. 2.23. At the centre of the cross-section of the rectangular waveguide (22.86×10.16 mm), a section of the waveguide-slot transmission line was located in the *E*-plane. The

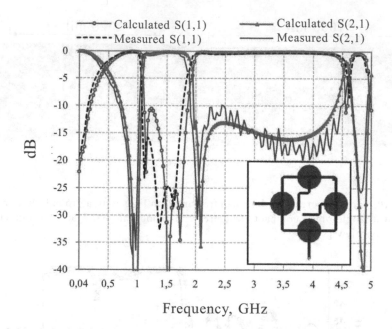

Fig. 2.21. Calculated and measured reflection and transmission spectra for a filter based on inwardly curved transmission lines and a structure with a ground plate containing a defect. On the inset – the filter circuit [15].

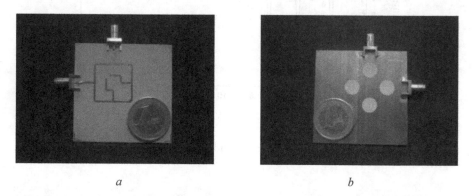

a   *b*

Fig. 2.22. Photos of created band-pass filters on ring resonators realized on the basis of structures with a ground plate containing a defect: *a* – top view, *b* – bottom view [15]

slot line is made on a ceramic (Al_2O_3, $\varepsilon = 9.6$) plate with a length of 23 mm, a width of 10.16 mm and a thickness of 1 mm. On one side of the plate was applied a metal coating thickness of 0.012 mm, the width of the gap in the coating was 4.0 mm. The segments of the waveguide-slot transmission line were separated by segments of a regular waveguide. This Bragg structure was investigated in the frequency range 8–12 GHz. The length of regular sections of the

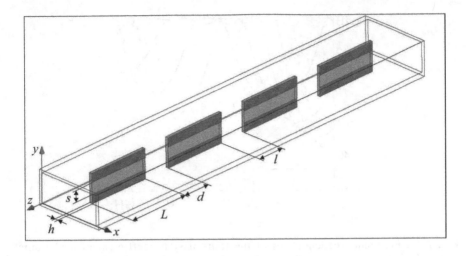

Fig. 2.23. Model of a photonic crystal based on waveguide-slot transmission lines: L is the length of a segment of the waveguide-slot transmission line; d is the length of the regular segment of the waveguide; h is the thickness of the substrate of the slot line, s is the slot width, l is the length of the central regular segment of the waveguide [16].

waveguide varied in the range from 2 to 10 mm. The violation of periodicity was created by changing the length of the central regular segment l of the waveguide in the range 14–20 mm.

It is known that in a periodic structure under certain relations between the parameters of the elements that form this structure, there are forbidden and allowed bands, while at frequencies, corresponding to the middle of the forbidden bands ω_{Bg}, the Bragg condition. For a structure composed of periodically repeating elements of two types, this condition can be represented in the form

$$\beta_1\left(\omega_{Bg}\right)d + \beta_2\left(\omega_{Bg}\right)L = n\pi, \tag{2.7}$$

where $d + L$ is the period of the structure with the photonic band gap, β_1 is the phase component of the wave propagation constant on the segment of the regular waveguide, which is a function of ω_{Bg}, β_2-phase component of the wave propagation constant on a segment of the waveguide-slot transmission line, determined by the geometric dimensions of the transmission slot and its electrical characteristics, n is the ordinal number of the forbidden band.

The change in the phase constants β_1 and β_2, as follows from the Bragg conditions, must lead to a shift of ω_{Bg}. The change in β_2 can be realized by changing the thickness and dielectric permittivity of

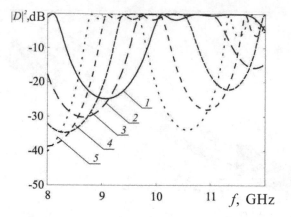

Fig. 2.24. Frequency dependences of the transmission coefficient of a photonic crystal for different substrate thicknesses h, mm: $1 - 0.2$; $2 - 0.3$; $3 - 0.4$; $4 - 0.5$; $5 - 0.6$. $L = 23$ mm; $d = 10$ mm; $s = 4.0$ mm; $\varepsilon = 9.6$ [16].

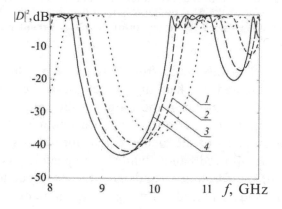

Fig. 2.25. Frequency dependences of the transmission coefficient of a photonic crystal for various permittivity of the substrate ε: $1 - 8$; $2 - 9$; $3 - 9.6$; $4 - 10$. $L = 23$ mm; $d = 10$ mm; $s = 4.0$ mm; $h = 1.0$ mm [16].

the substrate and changing the width of the slot [18]. The change in β_1 can be realized by changing the permittivity of the medium filling these segments

In [18] the results of computer simulation of the amplitude-frequency characteristics (AFC) of the microwave photonic crystal under consideration with the use of the HFSS system are presented. The results of calculating the frequency dependences of the squares of the moduli of the transmission coefficient $|D|^2$ of the microwave wave through the Bragg structure consisting of four segments of the waveguide-slot transmission line, for different thicknesses and

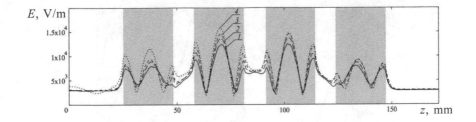

Fig. 2.26. Distribution of the electric field strength of an electromagnetic wave inside a photonic crystal along the direction of its propagation. h, mm: $1 - 0.2$; $2 - 0.3$; $3 - 0.5$; $4 - 0.6$; $x = 11.5$ mm; $s = 4.0$ mm. Gray areas are occupied by segments of the waveguide-slot transmission line, light regions are regular segments of the waveguide [16].

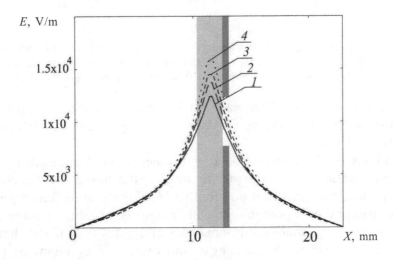

Fig. 2.27. Distribution of the electric field strength of an electromagnetic wave inside a photonic crystal in the transverse plane of the waveguide. h, mm: $1 - 0.2$; $2 - 0.3$; $3 - 0.5$; $4 - 0.6$; $z = 69.7$ mm; $s = 4.0$ mm [16].

dielectric permeability of the slot line substrate, are shown in Figs. 2.24 and 2.25, respectively.

From the results presented in these figures it follows that an increase in the thickness and dielectric constant of the slot line substrate within these limits leads to an increase in the phase constant β_2 of the segment of the waveguide-slot transmission line.

Figures 2.26 and 2.27 show the results of calculating the electric field strength of an electromagnetic wave inside a photonic crystal along the direction of its propagation (along the Z axis, see Fig. 2.23) in a plane passing through the middle of a wide waveguide wall and in the plane of the cross section of the waveguide passing through

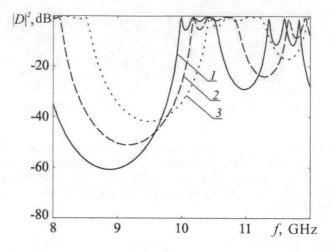

Fig. 2.28. Frequency dependences of the transmission coefficient $|D|^2$ of a photonic crystal for different widths of the slot of slot transmission line segments. s, mm: *1* – 2.0; *2* – 3.0; *3* – 4.0; $L = 23$ mm; $d = 10$ mm; $h = 1.0$ mm; $\varepsilon = 9.6$ [16].

the antinode of a standing wave on a segment of the waveguide-slot transmission line (along the X axis, see Fig. 2.23), for different substrate thicknesses.

Analysis of the distribution of the electric field strength of an electromagnetic wave inside a photonic crystal along the direction of its propagation makes it possible to conclude that at frequencies of transparency of a photonic crystal within its structure a standing wave mode with pronounced nodes and antinodes is realized there. With an increase in the thickness and dielectric constant of the substrate of the waveguide-slot transmission line, the intensity of the electric field in the antinodes increases, and decreases at the nodes. It should be noted that the maximum electric field strength in the plane of the cross section of the waveguide is located inside the middle of the substrate, but the character of the $E(x)$ dependence differs significantly from the sinusoidal characteristic of the H_{10} wave.

As noted above, the change in the width of the slot also leads to a change in the phase constant β_2 of the segment of the waveguide-slot transmission line, namely: with decreasing slot width, β_2 increases [18]. The results of the calculation of the frequency response of the investigated structure, shown in Fig. 2.28, indicate a shift in the frequency response to the low-frequency region with a decrease in the width of the gap.

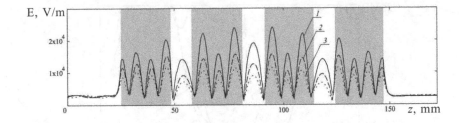

Fig. 2.29. Distribution of the electric field strength of an electromagnetic wave inside a photonic crystal along the direction of its propagation. s, mm: $1 - 2.0$; $2 - 3.0$; $3 - 4.0$; $x = 11.5$ mm; $h = 1.0$ mm. Gray areas are occupied by segments of the waveguide-slot transmission line, light regions are regular segments of the waveguide [16]

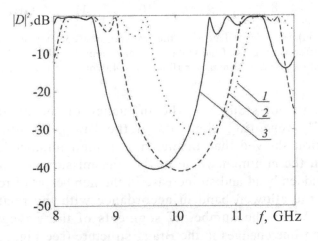

Fig. 2.30. Frequency dependences of the transmission coefficient $|D|^2$ of a photonic crystal for different lengths of regular waveguide sections d, mm: $1 - 2$; $2 - 5$; $3 - 10$. $L = 23$ mm; $s = 4.0$ mm; $h = 1.0$ mm; $\varepsilon = 9.6$ [16].

Figure 2.29 shows the results of calculating the electric field strength of an electromagnetic wave inside a photonic crystal along the direction of its propagation for a different width of the slot on a segment of the waveguide-slot transmission line. The analysis of the dependences shown in Fig. 2.29 shows that at the transparency frequencies of the photonic crystal, standing wave mode is realized within its structure, while the gap width in the waveguide-slot transmission line segment from 4.0 mm to 2.0 mm is doubled in antinodes. Additional studies have shown that increasing the lengths of regular segments of the waveguide d leads, as it follows from the Bragg relation, to the shift of the AFC structure to the low-frequency region (Fig. 2.30).

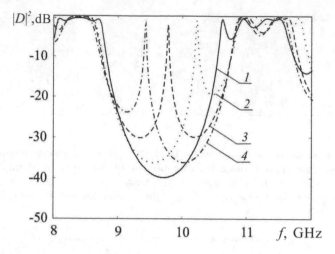

Fig. 2.31. Frequency dependences of the transmission coefficient of a photonic crystal for different lengths of the violated layer l, mm: $2 - 14$; $3 - 17$; $4 - 20$ and without violation $- 1$; $L = 23$ mm; $d = 10$ mm; $s = 4.0$ mm; $h = 1.0$ mm, $\varepsilon = 9.6$ [16].

The results of studies of the influence of the number N of periodically repeating elements in the Bragg structure under consideration showed that an increase in their number leads to a decrease in the minimum value of the transmission coefficient $|D|^2$ in the forbidden band and an increase in the number M of resonances that form the allowed band in accordance with the relation $M = N-1$ [19]. When the number of segments of the waveguide-slot transmission line changes in the Bragg structure (see Fig. 2.23) from 4 to 6, the transmission coefficient $|D|^2$ of the photonic crystal in the forbidden band decreases from -40 dB to -65 dB.

In order to elucidate the peculiarities of the appearance of an impurity (defect) mode resonance in the proposed Bragg structure, the results of investigations of the frequency response of a defective microwave photonic crystal with a changed length l of the central segment of a regular waveguide were carried out.

Figure 2.31 shows the results of calculating the frequency dependences of the transmission coefficient $|D|^2$ of a microwave wave through a Bragg structure consisting of four segments of a waveguide-slot transmission line with a length of 23 mm (curves $1-4$), with different violations l and without violation. The length of the disturbed layer l varied in the range from 14–20 mm.

From the results presented in Fig. 2.31 it follows that the presence of a violation of periodicity in the Bragg structure in the form of a modified length of the central segment of a regular waveguide leads to the appearance in the forbidden band of an impurity mode ('transparency window') whose position with increasing length of the violated layer *l* in the range of 14–20 mm is shifted to the low-frequency region.

In accordance with the model described above, the authors [16] created the Bragg structure. The experimental frequency response of the Bragg structure under consideration for different lengths of regular waveguide segments measured with the Agilent PNA-L Network Analyzer N5230A vector analyzer is shown in Fig. 2.32.

In [16], the results of investigations of the frequency response of the Bragg structure on the basis of segments of a waveguide-slot transmission line with a defect in the form of an altered length *l* of the central regular segment of the waveguide are presented. As follows from the experiment, the creation of a defect in a photonic crystal in the form of a central segment of a regular waveguide of increased length leads to the appearance of a transmission peak in the forbidden band and an increase in its width (curves *2* in Fig.

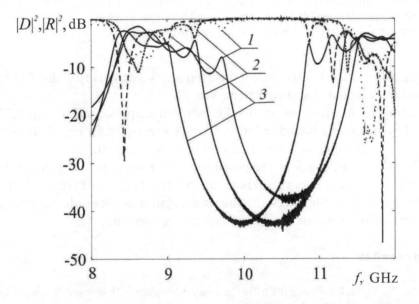

Fig. 2.32. Experimental frequency dependences of reflection coefficients $|R|^2$ (dashed, dotted and dash-dotted curves) and transmission coefficients $|D|^2$ (solid curves) of the Bragg structure for different lengths of regular waveguide segments *d*, mm: *1* – 2; *2* – 5; *3* – 10; $L = 23$ mm; $s = 4.0$ mm; $h = 1.0$ mm; $\varepsilon = 9.6$ [16].

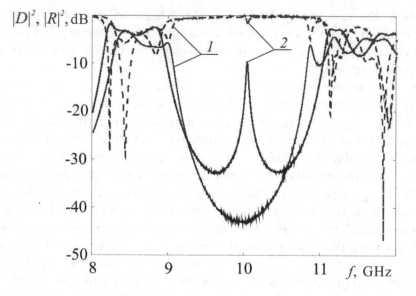

Fig.2.33. Experimental frequency dependences of the reflection coefficients $|R|^2$ (dashed curves) and the transmission coefficient $|D|^2$ (solid curves) of the Bragg structure in the presence of a violation in the form of a modified length of the central regular waveguide $l = 17$ mm (*2*) and without violation (*1*); $L = 23$ mm; $d = 10$ mm; $s = 4.0$ mm; $h = 1.0$ mm; $\varepsilon = 9.6$ [16].

2.33). Figure 2.33 also shows the frequency response of the Bragg structure without disturbance (curves *1*).

Comparison of the results of the experiment presented in Fig. 2.32 and Fig. 2.33 and the calculation results presented in Fig. 2.30 and Fig. 2.31 indicate their good quantitative agreement.

Thus, we can consider the possibility of creating a Bragg structure in the microwave range based on alternating segments of the waveguide-slot transmission line and regular waveguide segments theoretically grounded and experimentally proven.

References

1.   Tae-Yeoul, Kai Chang. IEEE Transactions on Microwave Theory and Techniques. – 2001. – Vol. 49, N. 3. P. 549–553.
2.   V.M. Mukhortov, S.I. Masychov, A.A. Mamatov, Vas. M. Mukhortov, Pis'ma v ZhTF 2013. Vol. 39, nNo. 20. P. 70–76.
3.   Md. Nurunnabi Mollaha, Nemai C. Karmakar, Jeffrey S. Fu, International Journal of Electronics and Communications (AEÜ), vol. 62, 2008, P. 717–724.
4.   Chul-Sik Kee, Jae-Eun Kim, Hae Yong Park, and H. Lim, IEEE Transactions on Microwave Theory and Techniques, Vol. 47, No. 11, November 1999. P. 2148–2150.

5.   B.A. Belyaev, A.M. Sergeantov, Radiotekhnika i Elektronika, 2005. Vol. 50, No. 8, p. 910–917.
6.   Nemai Chandra Karmakar and Mohammad Nurunnabi Mollah, IEEE Transactions on microwave theory and techniques, Vol. 51, No. 2, February 2003, P. 564–572.
7.   B.A. Belyaev, A.S. Voloshin, V.F. Shabanov, Dokl. Akad. Nauk, 2005, Vol. 403, No. 3, P. 319–324.
8.   D.A. Usanov, A.V. Skripal', A.V. Abramov et al., Zh. Tekh. Fiz. 2010. Vol. 80, No. 8, C. 143–148.
9.   D.A. Usanov, A.V. Skripal', A.V. Abramov, et al., Fizika volnovykh protsessov i radiotekhnicheskie sistemy. 2010. Vol. 13. No. 3. P. 26–34.
10.   Hideaki Kitahara, Tsuyoshi Kawaguchi, Junichi Miyashita, Mitsuo Wada Takeda, Journal of the Physical Society of Japan, April 15, 2003, Vol. 72, No. 4: P. 951–955.
11.   D. Nesic, Microwave Optical Tech. Letter, May 2003, Vol. 37, No. 5, P. 201-203.
12.   Sung-Il Kim, Mi-Young Jang, Chul-Sik Kee, et al., Current Applied Physics, 2005, No. 5, P. 619–624.
13.   Chul-Sik Kee, Mi-Young Jang, Sung-Il Kim, et al., Applied Physics Letters, 2005, Vol. 86, 181109.
14.   Yang F.-R., Ma K.-P., Qian Y., Itoh T., IEEE Transactions on Microwave Theory and Techniques. 1999. Vol. 47, issue 8. P. 1509–1514.
15.   El-Shaarawy H. B. Coccetti F., Plana R., et al, WSEAS Transactions on Communications, 2008. Vol. 7, No.11. P. 1112–1121.
16.   D.A. Usanov, S.A. Nikitov, A.V. Skripal', D.S. Ryazanov, Ratiotekhnika i elektronika. Vol. 61, No. 4. P. 1–6.
17.   V.G. Vinenko, D.A. Usanov, Microwave limiter. Author cert. USSR No. 1283878 MKI4 N01R 1/22. /. Publ. 01/15/1987. B.I. No. 2. P. 248.
18.   V.G. Vinenko, L.A. Fedoseeva, D.A. Usanov, Elektronnaya tekhnika. Ser. Elektronika SVCh. 1979. Vol. 3. P. 81–83
19.   D.A. Usanov, S.A. Nikitov, A.V. Skripal', D.V. Ponomarev, Ratiotekhnika i elektronika. 2013. Vol. 58. No. 11. P. 1071–1076.

3

Possibilities of controlling the characteristics of microwave photonic crystals by means of electric and magnetic fields

The ability to control the amplitude–frequency characteristics of microwave photonic crystals opens the prospect of expanding their field of application. This possibility was considered, in particular, in [1]. The authors of Ref. [1] took as their basis the design of the microwave photonic crystal used in Ref. 2 as a filter. This design was a template arranged in a rectangular waveguide in the form of a array of $n \times m$ metal disks (Fig. 3.1).

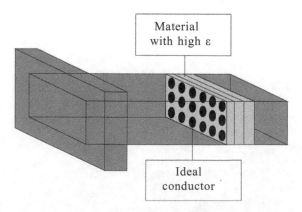

Material
with high ε

Ideal
conductor

Fig. 3.1. Filter on the basis of a printed metal–dielectric structure with a photonic band gap, placed in a rectangular waveguide [2].

The authors of [1] used templates of $n \times m = 1 \times 2$, 2×4, 3×6 metal disks in their experiments and templates in the form of a narrow metal strip located in the E-plane of the waveguide perpendicular to its wide wall. Ceramics (Al_2O_3) (1 mm thick) and polycrystalline iron–yttrium garnet (IYG) (1 mm thick) were used as the material forming the photonic crystal layers. Plates of these materials completely filled the waveguide along the cross section. The measurements were carried out in the frequency range 7.8–12.5 GHz. The thickness of the metallization layers was from 1 to 20 μm. To control the amplitude–frequency characteristics of the filter, an external magnetic field directed perpendicular to the wide wall of the waveguide was used. The results of the adjustment of the AFC filter consisting of 6 layers of an aluminum strip 0.5 mm wide deposited on a ceramic plate of the same width are illustrated in Fig. 3.2. The metal was located between the ferrite and Al_2O_3 ceramic plates. As follows from the results shown in Fig. 3.2, the authors of [1] succeeded in obtaining a filter bandwidth of ~0.5 GHz with a loss of ~6 dB and a loss in the locking band of ~40–43 dB. The corresponding dependences for metal templates and 2×4 disks are shown in Fig. 3.3.

Control of the frequency response in the structures under consideration occurs as a result of an increase in the real part of the magnetic susceptibility of ferrite with increasing magnetic field. This leads to an increase in the concentration of the microwave field in the ferrite and to an increase in the phase shift of the wave as it passes through the ferrite plate and, as a consequence, to the shift of the AFC.

The authors of [3] presented the results of a study of the behavior of a waveguide one-dimensional dielectric photonic structure with a ferrite plate as a defect. The frequency response was studied in

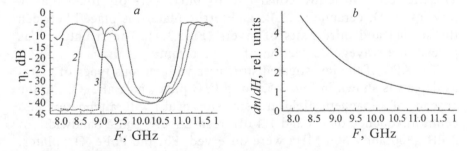

Fig. 3.2. Different rates of shift of the amplitude–frequency characteristics of the filter in an external field [1].

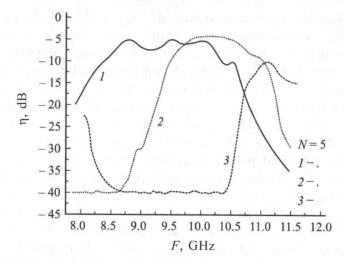

Fig. 3.3. Filter bandwidth modification based on a photonic bandgap structure under the influence of an external magnetic field H [1].

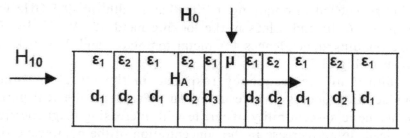

Fig. 3.4. One-dimensional dielectric photonic structure with a defect [3].

the frequency range 35–55 GHz. As a defect plates made from $BaFe_{12}O_{19}$ barium hexaferrite, from $PbFe_{12}O_{19}$ hexaferrite with a magnetoplumbite–ferritic structure with high uniaxial anisotropy and polycrystalline iron–yttrium ferrite garnet (IYG) were used. The dielectric structure consisted of four layers of 'foam plastic (ε_1, d_1) – Al_2O_3 ceramic (ε_2, d_2)'. The ferrite plate was placed between the second and third pairs of layers (Fig. 3.4). The structure was placed in a waveguide section of 5.2×2.6 mm^2.

The AFC of the investigated structures with various types of plates as defects is shown in Fig. 3.5. For a IYG plate and $BaFe_{12}O_{19}$, in the absence of a magnetic field, transmission peaks with attenuation at a central frequency of 1.8 and 1.4 dB, respectively, and a bandwidth of 3 dB ~300 and ~500 MHz were observed. For the $PbFe_{12}O_{19}$ plates, the attenuation value was ~10 dB for a much wider band.

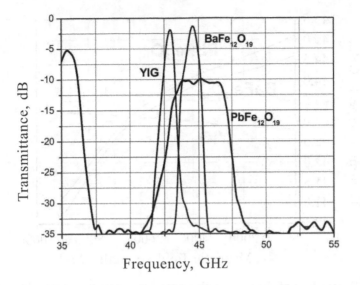

Fig. 3.5. Transmission peaks in the forbidden band for three ferrite plates [3].

The frequency of the transmission peak, as shown by the authors of [3], can be controlled by changing the magnitude of the external magnetic field. For ferrites with high anisotropy, the magnitude of the frequency shift in the centre of the transmission region toward the low frequencies was ~4 GHz with a change in the magnetic field by approximately 9000 Oe. The peak of the transmission bandwidth with the IYG plate slightly shifted towards higher frequencies (Fig. 3.6).

The authors of [4] used thin ferroelectric films with a dielectric constant of several thousand as the controlling element of a microwave photonic crystal. Thus, as the authors of [4] noted, the dielectric constant of films of composition $Ba_{0.8}Sr_{0.2}TiO_3$ can change by more than 3 times when a voltage of ~20 V is applied. The thickness of the films used by the authors of [4] was 40 nm. The microwave photonic crystal consisted of successively connected segments of a coplanar waveguide with wave resistances of 50 and 20 ohms. Waveguide conductors were applied to a heterostructure from a single-crystal MgO 0.5 mm thick and a $Ba_{0.8}Sr_{0.2}TiO_3$ film 40 nm in thickness. Forbidden bands were observed in the vicinity of 14 and 12 GHz frequencies, and a minimum of induced losses was observed in the vicinity of 28 GHz. Figure 2.4 shows the experimental dependences of the modul of the reflection coefficient S_{11} and the transmission coefficient S_{21} of the investigated photonic crystal obtained by the authors [4] in the absence of bias voltage (0 V) and with the application of a bias voltage of 40 V. It follows

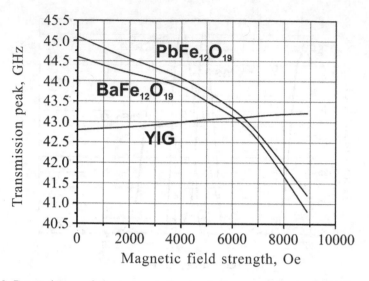

Fig.3.6. Dependence of the transmission peak frequency in the forbidden band of the photonic structure on the magnetic field for three ferrite grades [3]

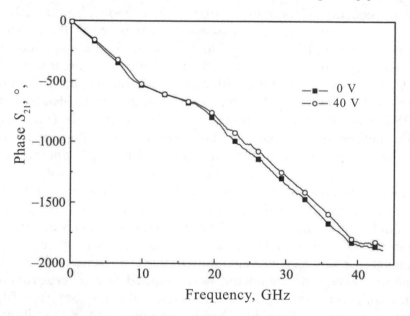

Fig. 3.7. Experimental frequency dependences of the phase of the transmission coefficient S_{21} of a photonic crystal obtained in the absence of a bias voltage (0 V) on a $Ba_{0.8}Sr_{0.2}TiO_3$ film and in case of applying a bias voltage of 40 V [4].

from the results in this figure that the central frequency of the photon forbidden band when the bias is applied shifts up the range by 0.75 GHz. Figure 3.7 shows the dependence of the S_{21} phase on the frequency at a bias voltage of 0 and 40 V, measured by the authors of Ref. [4]. It follows from the results that in the 18–38 GHz frequency band an electrical phase change of ~100° is possible.

The possibility of creating a waveguide photonic crystal with a tunable frequency position of the transparency window associated with a violation of periodicity in a photonic crystal and a decay value controlled by *p-i-n* diodes in this window is shown in [5]. An 11-layer microwave photonic crystal was designed to operate in the 3-cm wavelength range, consisting of 11 alternating layers of polycor ($\varepsilon = 9.6$) 1 mm thick and a foam ($\varepsilon = 1.1$) 12 mm thick. The violation of periodicity was ensured by using as a sixth layer a foam plate of reduced thickness. The calculated crystal transmission spectra for different values of the thickness of the sixth layer are shown in Fig. 3.8. Figure 3.9 shows the calculated frequency dependences of the modulus S_{21} and the phase *arg* S_{21} of the transmission coefficient for the thickness of the 6th layer of 3 mm.

To realize the control of the transmittance in the transparency band, we used a *p-i-n* diode array of type M34216-1 [6], which was incorporated into the waveguide path in conjunction with a photonic crystal (Fig. 3.10).

A control voltage regulated in the range 0–700 mV was applied to the array of *p-i-n*-diodes. The structure shown in Fig. 3.10 was placed in a waveguide of a 3-cm range and connected through a coaxial-waveguide transition to the 50-ohm coaxial path of the Agilent PNA-L network vector analyzer N5230A. With the help

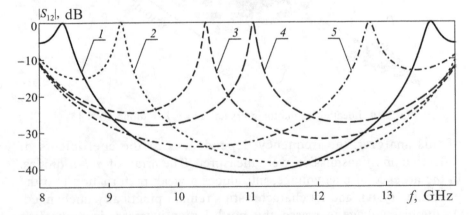

Fig. 3.8. Calculated dependences of the square of the modulus of the transmission coefficient of an 11-layer photonic crystal 'foam' for various values of the thickness of the disrupted 6th layer, d_6, mm: $2 - 7.0$, $3 - 4.0$, $4 - 3.0$, $5 - 1.0$. Curve *1* correspond to a photonic crystal without disturbances [5].

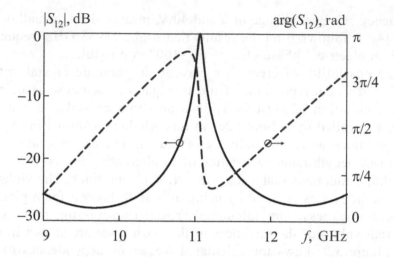

Fig. 3.9. Calculated frequency dependences of the modulus (solid line) and phase (dashed line) of the transmission coefficient of an 11-layer photonic crystal 'Al$_2$O$_3$ ceramic foam' for the thickness of the violated 6th layer $d_6 = 3$ mm [5].

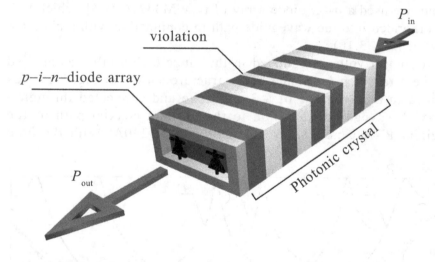

Fig. 3.10. Location of photonic crystal and *p–i–n*–diode array [5].

of this analyzer, the frequency dependences of the coefficients of reflection and transmission were measured. The array of *p–i–n*-diodes, in the absence of bias voltage, introduces a weak perturbation into the photonic crystal, and its characteristics remain practically unchanged. As the bias voltage increases, this perturbation increases in connection with the enhancement of the *i*-region by the charge carriers, and the resonant transmission characteristic of the photonic crystal decreases. Experimental frequency dependences $|S_{12}|$, *arg* S_{12} and $|S_{11}|$, *arg* S_{11} are shown in Fig. 3.11 for different values of bias voltage on the *p–i–n*-diode with the thickness of the disturbed layer d_6 = 5 mm. From the results shown in this figure it follows that the use of a microwave photonic crystal makes it possible to create a microwave switch with electrically adjustable characteristics from −1.5 to −25 dB when the bias voltage on the *p–i–n*-diodes varies from 0 to 700 mV.

As is known, the structures of the ring type possess properties of a photonic crystal, such as the presence of forbidden and allowed bands. The band nature of the spectrum in such structures is achieved due to multiple reflection from the inhomogeneity in the structure. Such devices in the microstrip form are described in [7,8]. In [9] the characteristics of such structures in the waveguide form are given. As an inhomogeneity, an element of the type 'metal pin with a gap' is used in them. This heterogeneity ensures the appearance of a resonant feature in the forbidden band of the system under investigation, called the defect mode of oscillation or the 'transparency window'.

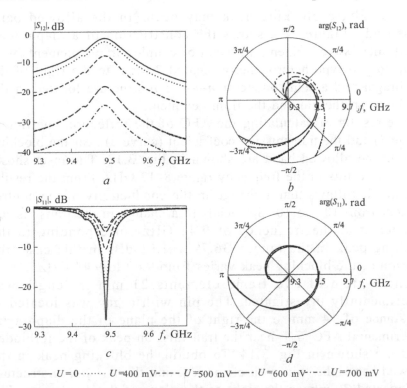

Fig. 3.11. Experimental dependences of the modulus (*a, b*) and phase (*b, d*) of the transmission (*a, b*) and reflection coefficients (*c, d*) of electromagnetic radiation in the region of the transparency window of a photonic crystal for various values of the *p–i–n*-diode voltage, d_6 = 5.5 mm [5].

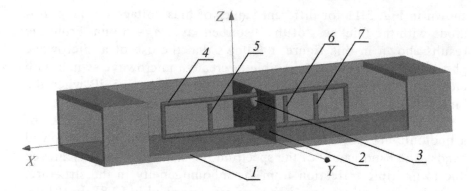

Fig. 3.12. Model of the microwave element based on the diaphragm and the system of coupled frame elements containing heterogeneity in the form of 'a pin with a gap' designs: *1* – waveguide, *2* – diaphragm, *3* – hole, *4* – frame element, *5–7* – inhomogeneities of the 'pin with a clearance' type [9].

Accordingly, a blocking peak may occur in the allowed band (passband). Figure 3.12 shows the construction of a diaphragm-based microwave element, a system of coupled frame elements with a gap-to-gap type heterogeneity located 20 mm to the right of the diaphragm and a semiconductor *n–i–p–i–n* structure located in the gap between the pin and the frame element.

The results of calculating the AFC of the reflection coefficient (curve *1*) and the transmission coefficient (curve *2*) near the blocking peak in the allowed band are shown in Fig. 3.13. The inset shows the same results in the frequency range 8–12 GHz. From the results obtained it follows that a change in the conductivity of the control element from 10^{-3} to 10^5 S/m leads to a change in the transmission coefficient at the frequency of 9.44 GHz corresponding to the blocking peak in the range $-36.79...-1.01$ dB. In this case, the position of the blocking peak varies from 9.69 to 9.44 GHz.

The system of two frame elements 21 mm in length was experimentally investigated. The pin with a gap was located at a distance of 14 mm to the right of the plane of the diaphragm. Experimental AFC $|S_{11}|^2$ near the transmission peak of the forbidden band are shown in Fig. 3.14. To obtain the blocking peak in the allowed band, the 'pin with clearance' and the *n–i–p–i–n*-structure was located 20 mm to the right of the plane of the diaphragm. The bias voltage applied in the range 0–9 V resulted in a change in the transmission coefficient from -25 to 1.5 dB at 9.644 GHz, with the peak position varying from 10.079 to 9.644 GHz.

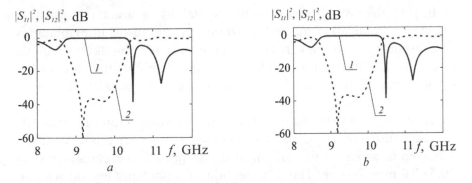

Fig. 3.13. The calculated frequency dependences of the reflection coefficient (curves *1*) and the transmission coefficient (curves *2*) of the microwave device on the basis of the diaphragm and the system of coupled frame elements: *a* – not containing inhomogeneity, *b* – containing 'pin with a gap' inhomogeneities [9].

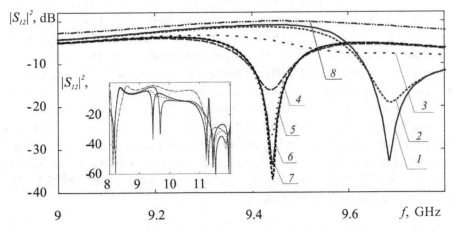

Fig. 3.14. Calculated amplitude–frequency characteristics of the transmission coefficient near the blocking peak of the allowed band of the microwave element with a non-uniformity of the 'pin with a gap' type and n–i–p–i–n-structure. σ, S/m: *1* – 10^{-3}, *2* – 1.0, *3* – 10.0, *4* – 10^2, *5* – 10^3, *6* – 10^4, *7* – 10^5, *8* – the case of the absence of the pin and the control element [9].

It is known that resonant waveguide diaphragms are often used elements of the construction of attenuators and switches on p–i–n-diodes. The small geometric dimensions of the slot provide an effective interaction between semiconductor elements having small dimensions with the field of the waveguide, and the use of a resonant diaphragm in measuring systems makes it possible to increase the locality of the measurements [10], therefore consideration of the properties of photonic crystals on waveguide resonance diaphragms is topical.

In [11], the investigated photonic crystal was a structure consisting of seven periodically arranged rectangular metal resonant diaphragms at a distance $L = 20$ mm from each other in a rectangular waveguide of a three-centimeter range. The width and height of the slots of the diaphragms of the photonic crystal were chosen equal to 20 and 2 mm, respectively.

To control the resonance properties of such photonic crystals, we used an $n–i–p–i–n$ diode array consisting of four diode elements arranged in a central diaphragm in the form of two rectangular slots 10.5×1.0 mm² in size. The construction of a photonic crystal with an $n–i–p–i–n$-diode array, acting as a disturbance with current-controlled characteristics, is shown in Fig. 3.15 *a*.

When calculating, it was assumed that the semiconductor array consists of $n–i–p–i–n$-structures having the form of a parallelepiped of height $h = 1$ mm and a cross section of 1.0×0.5 mm² (Fig. 3.15 *b*).

On the basis of numerical modelling using the finite element method in the ANSYS HFSS program, the amplitude–frequency characteristics of the reflection and transmission coefficients of a photonic crystal under different specific electrical conductivities of the i-layer of the $n–i–p–i–n$ structure were investigated. It was assumed that for a forward bias, the specific electric conductivity σ of a given element varied in the range $10^{-2}...10^{5}$ S/m. Such a change in the electrical conductivity value σ due to the enhancement of the i-regions by the injected charge carriers corresponds to a change in the current flowing through the $n–i–p–i–n$ structure of the 2A505 type in the range $0...300$ mA.

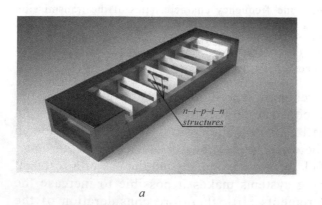

a   *b*

Fig. 3.15. Constructions of: *a* – photonic crystal with an $n–i–p–i–n$-diode array, *b* – $n–i–p–i–n$-structure (*b*) [11].

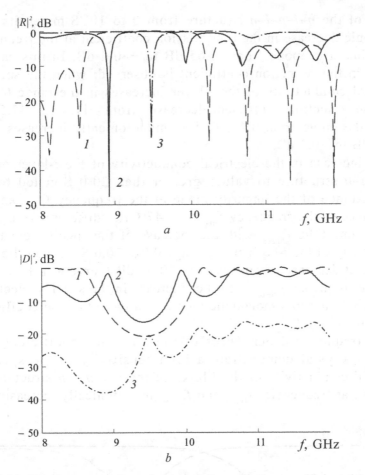

Fig. 3.16. The frequency dependences of the reflection coefficient (*a*) and the transmission coefficient (*b*) of a photonic crystal without violations consisting of seven diaphragms without a $n-i-p-i-n$-array (curve *1*) and with a control $n-i-p-i-n$-array as a violation of the central layer at a specific the electrical conductivity of the i-layer of the $n-i-p-i-n$ structure is $\sigma = 0.4$ (curve *2*) and 1000 S/m (curve *3*) [11].

The results of calculations of the amplitude–frequency characteristics of a photonic crystal are shown in Fig. 3.16.

As follows from the calculation results, a photonic crystal without violations created on the basis of resonant diaphragms is characterized by the presence of a forbidden band in the frequency range 8.53...10.23 GHz (curves *1* in Fig. 3.16, *a*). The introduction of a $n-i-p-i-n$-array into the photonic crystal as a violation of the central layer leads to the appearance of an impurity mode resonance in the forbidden band of a photonic crystal at a frequency $f_{1\text{theor}} = 8.91$ GHz. A change in the electrical conductivity of the *i*-layer of the *n–i–p–i–n* structure from 0 to 10^4 S/m results in a monotonic decrease in the transmission coefficient at the frequency of the impurity mode from −0.65 dB to −40.6 dB. In this case, a change in the reflection coefficient is observed: when the specific electrical conductivity of the *i*-layer increases in the range 0...0.4 S/m, the reflection coefficient decreases from −11.3 to −47.5 dB, and in the range from 0.4...10^4 S/m monotonically increases from −47.5 dB to −0.3 dB.

The increase in the electrical conductivity of the *i*-layer of the *n–i–p–i–n* structure to values greater than 20.0 S/m led to the disappearance of the impurity mode at the frequency $f_{1\text{theor}}$ and its appearance at the frequency $f_{2\text{theor}}$ = 9.47 GHz, different from $f_{1\text{theor}}$. At the same time $f_{2\text{theor}}$, with the increase of the specific electrical conductivity of the *i*-layer in the range 0.0...270.0 S/m, the reflection coefficient decreased from −0.1 to −30.66 dB (curve *3* in Fig. 3.16, *a*) at the frequency $f_{2\text{theor}}$. A further increase in the specific electrical conductivity led to a monotonic increase in the reflection coefficient at the frequency $f_{2\text{theor}}$.

The frequency dependences of the reflection coefficient of a photonic crystal demonstrate a high sensitivity to the specific electrical conductivity of the *i*-layer of the *n–i–p–i–n* structure. In this case, at frequencies $f_{1\text{theor}}$ and $f_{2\text{theor}}$, monotonically increasing or

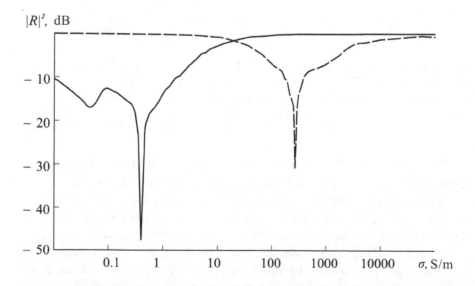

Fig. 3.17. Dependences of the reflection coefficient on the value of the electrical conductivity of the *i*-layer of the *n–i–p–i–n* structure at the frequencies of the impurity modes of the photonic crystal $f_{1\text{theor}}$ = 8.91 GHz and $f_{2\text{theor}}$ = 9.47 GHz [11].

monotonically decreasing as well as non-monotonic dependences of the reflection coefficient on the value of the electrical conductivity of the *i*-layer can be obtained if its range is varied (Fig. 3.17).

The disappearance of the impurity vibration mode at the frequency f_{1theor} and its appearance at the frequency f_{2theor} with an increase in the specific electrical conductivity of the *i*-layer of the *n–i–p–i–n* structure up to a certain value is due to the effect of a change in the type of resonant reflection of electromagnetic radiation from the so-called half-wave resonance to the quarter wave [12].

In the photonic crystal described above, the *n–i–p–i–n*-array plays the role of the conducting layer on resonant diaphragms. With a small thickness and low conductivity of the conductive layer at a frequency f_{1teor} corresponding to the minimum of the reflection coefficient, the distribution of the electric field in the standing microwave wave is realized, at which an antinode is formed at the boundary of the conducting layer, and at a large thickness and high conductivity, the minimum reflection coefficient arises at a frequency f_{2theor}, which is different from the frequency f_{1theor}, which is caused by the appearance of a new electric field distribution in a photonic crystal, on the border of the conductive layer is formed of an electric field of an electromagnetic wave node.

The photonic crystal constructed in accordance with the model described above and consisting of seven aluminium resonant diaphragms 10 µm in thickness was studied experimentally, the distance between them was chosen equal to 20 mm and fixed with a layer of foam completely filling the cross section of the rectangular waveguide. In a photonic crystal in the central diaphragm, an *n–i–p–i–n*-array of type 2A505 diodes was placed as a disturbance with controllable characteristics (see Fig. 3.15).

The amplitude–frequency characteristics of the transmittance and reflection coefficients of the investigated photonic crystals were measured with the Agilent Microwave Network Analyzer N5242A PNA-X network analyzer in the frequency range 8...12 GHz.

The choice of the size of the aperture slots in accordance with the model of the photonic crystal described above ensured the emergence of one allowed and one forbidden band in the frequency range 8...12 GHz on the frequency dependences of the transmission coefficients $|D|^2$ and the reflection $|R|^2$. Figure 3.18 shows the results of AFC measurements of the reflection and transmission coefficients of a photonic crystal without violation of the periodicity and with the

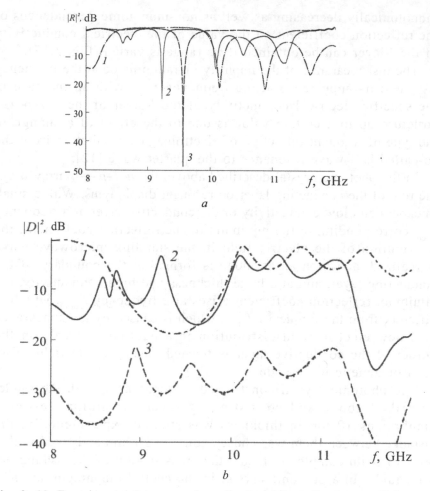

Fig. 3 .18. Experimental frequency dependences of the reflection coefficient (*a*) and the transmission coefficient (*b*) of a photonic crystal without disturbances consisting of seven diaphragms without an *n–i–p–i–n* array (curve *1*) and with a control *n–i–p–i–n* array as a violation of the central layer at the values of the control current $I = 0.0005$ (curve *2*) and 8.15 mA (curve *3*) [11].

n–i–p–i–n-array introduced into the photonic crystal as a violation of the central layer with controlled characteristics.

As follows from the results of the experiment, a photonic crystal without violations, created on the basis of resonant diaphragms, is characterized by the presence of a forbidden band in the frequency range 8.14...10.14 GHz (see Fig. 3.18, *a*, curve *1*). The introduction of an *n–i–p–i–n*-array into the photonic crystal as a violation of the central layer leads to the appearance of an impurity mode resonance in the forbidden band of a photonic crystal at a frequency $f_{1exp} = 9.22$

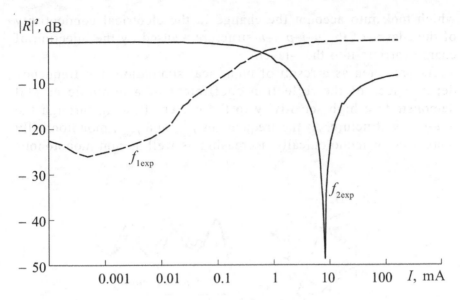

Fig. 3.19. Dependence of the reflection coefficient on the current flowing through the n–i–p–i–n-structure at the frequencies of impurity modes of the photonic crystal $f_{1exp} = 9.22$ GHz and $f_{2exp} = 9.56$ GHz [11].

GHz. A change in the magnitude of the control current of the n–i–p–i–n-structure from 0 to 0.5 μA leads to a decrease in the reflection coefficient at the frequency of the impurity mode of oscillations f_{1exp} from −22.5 to −25.2 dB. A further increase in the control current results in a monotonic increase in the reflection coefficient at a frequency f_{1exp} to −0.61 dB. The transmission coefficient in the entire range of control currents decreased monotonically from −3.42 to −35.0 dB.

An increase in the current flowing through the n–i–p–i–n-structure to values greater than 1.0 mA led to the disappearance of the impurity mode at the frequency f_{1exp} and its appearance at the frequency $f_{2exp} = 9.56$ GHz, different from f_{1exp}. At the same time, the reflection coefficient decreased from −0.1 to −48.64 dB (Fig. 3.19) at the frequency f_{2exp} with increasing current flow through the n–i–p–i–n-structure in the range from 0.0 to 8.15 mA. A further increase in the control current resulted in a monotonic increase in the reflection coefficient at a frequency f_{2exp}.

This behaviour of the amplitude–frequency characteristic of the reflection and transmission coefficients with increasing current flowing through the n–i–p–i–n-structure is in good agreement with the results of numerical simulation of the investigated photonic crystal, which took into account the change in the electrical conductivity of the *i*-layer of the *n–i–p–i–n*-structure, caused by the injection of charge carriers into the *i*-layer.

As predicted as a result of numerical simulation, the frequency dependences of the reflection coefficient of a photonic crystal demonstrate a high sensitivity to the current flowing through the *n–i–p–i–n*-structure. At the frequencies $f_{1\text{exp}}$ and $f_{2\text{exp}}$, monotonically increasing or monotonically decreasing as well as non-monotonous

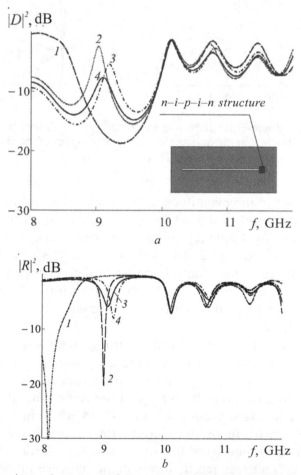

Fig. 3.20. Experimental response of the transmission (*a*) and reflection coefficients (*b*) of a photonic crystal at resonant diaphragms without disturbance (curves *1*, aperture size a = 20 mm, b = 2 mm, constant distance L = 20 mm) and with electrically controlled gap size (a = 15 mm, b = 0.3 mm) of the central diaphragm, which acts as a violation, for various values of the forward current flowing through the *n–i–p–i–n*-structure: I = 0.0 (*2*), 4.0 (*3*), 550 mA (*4*) [11].

dependences of the reflection coefficient on the current flowing through the $n–i–p–i–n$-structure were obtained (see Fig. 3.19).

Thus, as follows from the results of the experiment, the reflection coefficient at the frequency of the impurity mode of the oscillations f_{1exp} varied from −25.2 to −0.6 dB, and at the frequency of the impurity mode f_{2exp} from −0.1 to −48.64 dB.

The obtained experimental results are in good agreement with the results of numerical calculations and indicate the possibility of creating an electrically controlled modulator and a microwave signal switch in a reflection scheme on the basis of a photonic crystal at resonant diaphragms. In this case, at the frequency f_{1exp} it is possible to implement a direct switching mode, i.e. increase in the reflection coefficient when current is passed through the $n–i–p–i–n$-structure, and at the frequency f_{2exp} of the direct and inverse.

Calculations and experimental studies show that in photonic crystals on resonant diaphragms made of metal foil, to create an impurity mode resonance in the forbidden band, it is necessary to change the distance between the selected diaphragms or to change the width of the diaphragm slot. When using a photonic crystal with a modified slot width of one of the diaphragms, for example, a central one, the frequency position of the impurity mode is determined by the width of the slot of this diaphragm and, as it decreases, shifts toward the high-frequency edge of the forbidden band.

To create a photonic crystal on resonant diaphragms with an electrically controllable size of the diaphragm slot, which serves as a disruption, a structure can be used in which a $n–i–p–i–n$-structure is located near one of the edges of the central diaphragm with a slot of a reduced width (see the inset in Fig. 3.20 a).

When the direct current flows through the $n–i–p–i–n$-diode structure, the i-layer is enhanced with free carriers, the $n–i–p–i–n$-diode structure acts as a conductive inclusion, which decreases the width of the diaphragm slot and at large direct currents shifting the position of the impurity mode resonance toward the high-frequency edge of the forbidden band.

Figure 3.20 shows the frequency response of the transmission and reflection coefficients of a photonic crystal at resonant diaphragms without disturbance and with an electrically controlled gap size of the diaphragm, which acts as a disturbance.

As follows from the results of the experiment, when a central diaphragm with an electrically controlled size acts as a disturbance in the structure of a photonic crystal on resonance diaphragms, an impurity mode resonance appears at the frequency f_{1exp} = 9.03 GHz in the forbidden band of a photonic crystal, which shifts in frequency by an amount equal to 160.0 MHz, in the direction of the high-frequency edge of the band gap with a forward current of 550 mA passing through the *n–i–p–i–n*-diode structure. At that, an increase in the reflection coefficient from −20.35 dB at I = 0 mA to −1.65 dB at I = 550 mA is observed at the frequency of the impurity mode of 9.03 GHz and the transmission coefficient decreases from −4.0 dB at I = 0 mA to −16.0 dB at I = 550 mA.

It should be noted that the computer simulation of the characteristics of a photonic crystal with an electrically controlled gap size of the central diaphragm, which acts as a disturbance, shows the possibility of a significant reduction in the direct losses when the specific conductivity of the *i*-layer reaches 10^5 S/m. Such an increase in the electrical conductivity of the *i*-layer can be achieved by using *n–i–p–i–n*-structures with heterojunctions that provide a higher level of injection than the traditional homojunctions.

The results obtained by the authors of [11] testify to the existence of impurity mode resonance at two frequencies in the forbidden band of a photonic crystal when the current in the *n–i–p–i–n*-diode array changes, acting as a conductive layer in a photonic crystal at resonant diaphragms. The possibility of realization of a photonic crystal at resonant diaphragms is established, in which the size of the central diaphragm, performing the role of disturbance, is electrically controlled by the *n–i–p–i–n*-diode structure. The possibility of practical realization of an electrically controlled modulator and a microwave switch operating in both direct and inverse modes with a dynamic range of more than 45 dB based on a photonic crystal at resonant diaphragms is shown.

References

1. Britun N.V., Danilov V.V., Pis'ma Zh. Tekh. Fiz., Vol. 29, No. 7. P. 27–32.
2. Kuriazidou C.A., Contopanagos H. F., Alexopolos N.G., IEEE Transactions on microwave theory and techniques. 2001. V. 49. N. 2. P. 297–306.
3. Danilov V.V., Launets V.L., Oleinik V.V., in: International Crimean conference 'Microwave engineering and telecommunication technologies': in 2 volumes - Sevastopol: Weber, 2007, Vol. 2. - P. 558–569.
4. Mukhortov V.M., Masichev S.I., Mamatov A.A., Mukhortov Vas.M., Pis'ma Zh. Tekh. Fiz. 2013. Vol. 39, No. 20. P. 70–76.
5. Usanov D.A., Skripal' A.V., Abramov A.V., et al., Izv. VUZ. Elektronika. 2010. No. 1. P. 24–29.
6. M-34216-1 X-range switch. - URL: http: //www.oao-tantal.ru/tovar.php?id=3394

7. Sung-Il Kim, Mi-Young Jang, Chul-Sik Kee, et al., Current Applied Physics, 2005, No. 5, P. 619–624.
8. Chul-Sik Kee, Mi-Young Jang, Sung-Il Kim, et al., Applied Physics Letters, 2005, vol. 86, 181109.
9. Usanov D.A., Nikitov S.A., Skripal' A.V., et al., Radiotekhnika i elektronika. 2014, Vol. 59. No. 11. P. 1079–1084.
10. Usanov D.A., Gorbatov S.S., Kvasko V.Yu., Fadeev A.V. Pribory i tekhnika eksperimenta. 2015. No. 2. P. 77–83.
11. Usanov D.A., Nikitov S.A., Skripal A.V., et al., Journal of Communications Technology and Electronics, 2018, Vol. 63, No. 1, P. 58–63
12. Mukhortov V.M.,et al., Pis'ma Zh. Tekh. Fiz., 2013. Vol. 39, No. 20, P. 70–76.

4

Applications of microwave photonic crystals

Different directions of practical use of the unique characteristics of microwave photonic crystals were considered in [1]. The authors of this work observed almost complete transmission of electromagnetic waves through bending in a flat hollow waveguide. An increase in the detected signal was shown when the detector was placed in the region of violation of the periodicity in a microwave photonic crystal. A similar defect was used to increase the directivity of the antennas. As periodic gratings, the authors of [1] considered structures of the 'layer on the layer' type (Fig. 4.1 a). This grating consisted of aluminum rods ($0.32 \times 0.32 \times 15.25$ cm). Figure 4.1 b shows the scheme for measuring the transmission characteristics of an electromagnetic wave through such a structure, Fig. 4.1 c shows the transmission and reflection characteristics for such a structure. The photonic band gap was observed for three different directions of the wave incidence in the range from 10.6 to 12.7 GHz. When the periodicity was violated, decaying (higher) types of waves appeared in the system. Figure 4.1 c shows the transfer spectrum of a 16-layer crystal in [1] with a single rod absent in the 8th layer. The resonant frequency of the defect mode was 12.16 GHz (Fig. 4.1 d). The defective mode with a change in the width of the air gap to 8.6 mm was detected at the frequency 11.61 GHz.

Figure 4.2 a schematically shows an installation for measuring the characteristics of a wave passing through a photonic crystal. Figure 4.2 b shows the change in the transmission bandwidth of a photonic crystal for different values of the air gap width, which

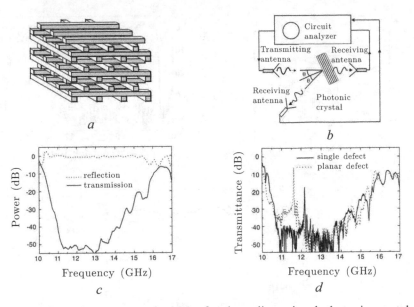

Fig. 4.1. Schematic representation (*a*) of a three-dimensional photonic crystal of the 'layer on a layer' type; *b* – experimental setup for measuring the transmission and reflection spectra of a photonic crystal; *c*) transmission pattern (solid line) and reflections (dashed line) of the periodic structure consisting of 4 cells along the packing direction; *d* – the transmission spectra of structures with a single (solid line) and planar (dashed line) defects [1].

plays the role of a defect. The authors of [1] noted almost 100% propagation of a wave through a crystal in a certain frequency range. Figure 4.2 *a* shows a construction characteristic of the *L*-bending of the waveguide. Figure 4.2 *c* shows the frequency dependence of the decay as the wave passes through the bend. The maximum amplitude of the transmitted signal was 35% of the amplitude of the incident signal. The authors of [1] presented the results of measurements of wave propagation through the structures of a 3D photonic crystal with strongly localized defects (Fig. 4.3). The measurement scheme is given in Fig. 4.4. The defect of the 'strip' was created by removing one rod from each cell, as shown in Fig. 4.4. For this type of defect, almost complete wave propagation was observed in the frequency range 11.47–12.62 GHz (see Fig. 4.3 *a*). Full transmittance in a certain frequency range was also observed with a zigzag arrangement of defects.

As an application of the properties of microwave photonic crystals, the authors of Ref. 1 cite an example of improving the characteristics of defects when they are placed in a resonance cavity formed by a grating defect. The receiving part of the detector was the open end of

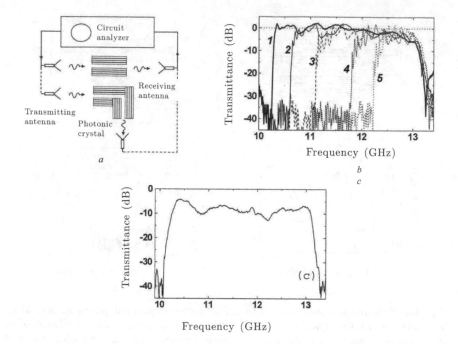

Fig. 4.2. Experimental setup (*a*) for investigation of waveguide structures with parallel plates (above) and *L*-bend (below); *b* – transmission spectra of waveguide structures with parallel plates with a change in the width of the gap, mm: 1–18, 2–16, 3–14, 4–12 and 5–10.5. *c* is the transmission spectrum for the *L*-bending of the waveguide [1].

the coaxial cable. At the defect frequency of 11.68 GHz, an increase in the power input to the detector was observed 1600 times. The photonic crystal on each side of the cavity was considered by the authors of [1] as a Fabry–Perot resonator. The amplification factor of the microwave detector signal due to the circuit used was 450. The bandwidth of the detector was within 10.5–12.8 GHz. The authors [1] described the use of a 3D photonic crystal in the antenna design. The antenna was mounted on a substrate of a photonic crystal to achieve a high directivity of its radiation. This antenna contained a radiation source in combination with a resonator based on a photonic crystal of the 'layer on layer' type. The photonic crystal used had 20 layers. The defect was formed by the appropriate choice of the 8th and 9th layers of the structure. The measured half-width along the H-plane was 12° and along the *E*-plane 11° (Fig. 4.5). The authors of [1] report the possibility of adjusting the resonant frequency of radiation in the considered case in the range from 10.6 to 12.8 GHz.

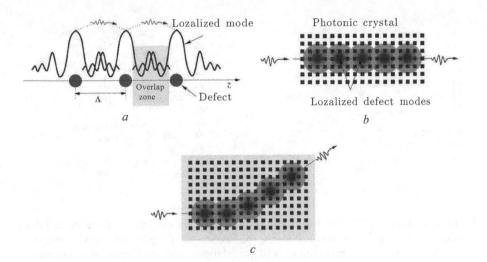

Fig. 4.3. Scheme of propagation of photons (*a*) by means of jumps between coupled decaying defect modes. The overlap of the defect modes is large enough to ensure the propagation of electromagnetic waves along strongly coupled resonant modes; *b* – the mechanism of light propagation through localized defect modes in a photonic crystal. *c* – bending of electromagnetic waves around sharp angles [1].

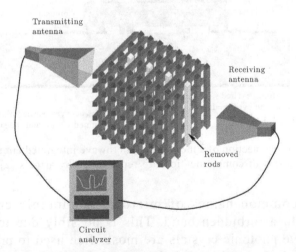

Fig. 4.4. Experimental setup for measuring the amplitude–frequency and phase-frequency dependences of the transmission coefficient for waveguides on coupled resonators in three-dimensional 'layer on the layer photonic crystals' [1].

In [2], the structure of a microwave photonic crystal made on the basis of a coplanar line (Fig. 4.6) was used to suppress unwanted waves of higher types and thereby improve the performance characteristics of a microwave amplifier of the millimeter wavelength range.

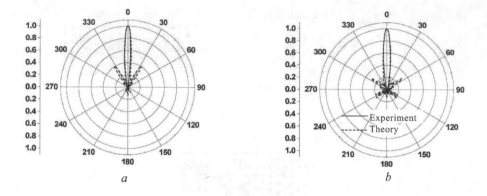

a   *b*

Fig. 4.5. Measured (solid lines) and calculated (dashed lines) of the directivity pattern of a single-pole antenna inside the resonator of a photonic crystal for: *a* – *H*-field, *b* – *E*-field. Measurements and calculations were performed at a resonance frequency of 11.7 GHz [1].

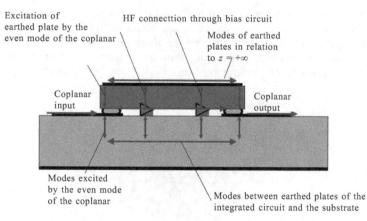

Fig. 4.6. Scheme of connection of a monolithic microwave integrated circuit. Possible ways and mechanisms of communication are represented by arrows [2].

One of the common names of microwave photonic crystals is structures with a forbidden band. This is possibly due to the fact that microwave photonic crystals are most often used to provide the stop band. At the same time, it is known that, in addition to the stop band (an analog of the band gap in semiconductors), the microwave frequencies of photonic crystals have a frequency band in which the wave propagates practically without decaying. The authors of [3] proposed to use this property to create waveguide broadband matched loads. The problem of constructing such loads remains one of their topical problems in microwave electronics at the present time. Matched microwave loads are widely used both independently and as elements of complex functional devices: directional couplers, adders, power meters, measuring bridges, microwave filters, etc. [4, 5].

One of the main tasks to be solved when creating microwave matched loads, designed for operation at small and medium power levels of microwave radiation, is to ensure matching in the widest possible frequency range with minimum dimensions.

The results of studies of the frequency dependences of the reflection coefficients of electromagnetic radiation from one-dimensional photonic crystals containing, along with periodically changing dielectric filling, nanometer metal layers are presented in [6, 7]. It was shown that the reflection coefficient in such systems can vary within very wide limits with insignificant variations in the thickness of the metal film not exceeding several tens of nanometers.

To determine the possibility of using the one-dimensional photonic crystals, computer simulations were carried out as matched loads. As the investigated microwave-photonic crystal, we used a structure, a schematic image of which is shown in Fig. 1.5.

When calculating the reflection coefficients R and the transmittance coefficien T of the electromagnetic wave, a wave transfer matrix was used between the regions with different values of the transmittance constant of the electromagnetic wave, as was described in detail in [7–9]. In the course of computer modelling, in [3] the possibility of creating matched waveguide loads in 8 mm and 3 cm wavelength bands was demonstrated. The investigated load was a photonic crystal based on a multilayer metal–dielectric structure with various thicknesses, permittivity, and electrical conductivity of the layers, which was placed in the waveguide and completely filled it along the cross section, the plane of the layers being perpendicular to the direction of propagation of the electromagnetic wave. The number of layers, their thickness, dielectric permeability, electrical conductivity, and the order of alternation of layers were determined as a result of solving the optimization problem in such a way that the values of reflection and transmission coefficients were less than the set values in the chosen frequency range. The optimization was carried out numerically using the Levenberg–Marquardt iterative method. The results of calculating the squares of the moduli of reflection coefficients $|R|^2$ (solid curve) and the $|T|^2$ (dashed curve) transmittance coefficients of an electromagnetic wave in the 8–12 GHz band with its normal incidence on a fully filling waveguide along the cross section of a multilayer metal–dielectric structure consisting of 6 alternating layers of nanometer metal film (chromium) deposited on a ceramic substrate (Al_2O_3) are presented in Fig. 4.7. As follows from the calculation results, the use of the proposed photonic crystals with nanometer metal layers makes it possible to provide the magnitude of the square of the modulus of reflection coefficient $|R|^2$ of electromagnetic radiation in the range 8–12 GHz of less than 1.0%. In this case, the square of the modulus of the transmittance $|T|^2$ does not exceed 0.32%. Figure 4.8 shows the results of calculating the frequency dependences of $|R|^2$ (solid curve) and $|T|^2$ (dashed curve) in the range 28–40 GHz when a multilayer metal–dielectric structure consisting of 13 alternating layers of a nanometer metallic film (chrome) deposited on a waveguide ceramic substrate (Al_2O_3). The results of the calculation show the possibility of realizing broadband matched loads in the 8-mm wavelength range using photonic crystals with nanometer metal layers. The values of $|R|^2$ and $|T|^2$ in the range 31–36 GHz do not exceed 1.0%.

In the experimental realization of microwave photonic crystals intended for use in the three-centimeter range of wavelengths, metallic layers (chromium) were deposited on Al_2O_3 ceramic substrates with a permittivity $\varepsilon = 9.6$. In addition, we used fluoroplastic layers with a dielectric permittivity $\varepsilon = 2.1$. Layers tightly pressed

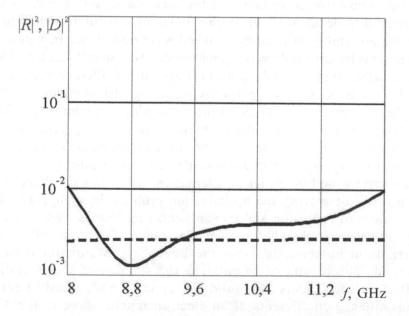

Fig. 4.7. Frequency dependences of $|R|^2$ (——) and $|T|^2$ (– – –) for the matched load in the three-centimeter range of wavelengths [3].

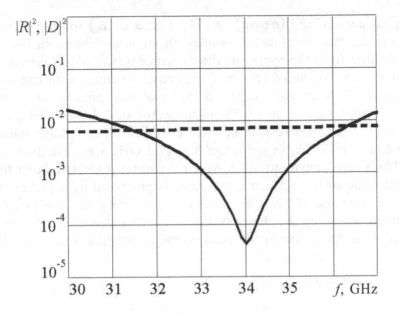

Fig. 4.8. Frequency dependences of $|R|^2$ (——) and $|T|^2$ (– – –) for the matched load in the eight-millimeter wavelength range [3].

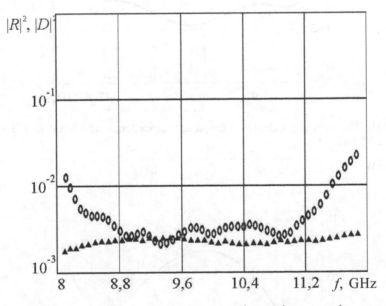

Fig. 4.9. Experimental frequency dependences $|R|^2$ (oooo) and $|D|^2$ (▵ ▵ ▵ ▵) for the matched load in the three-centimeter range of wavelengths [3].

against each other mechanically and placed in a waveguide. The measurements were carried out using a panoramic VSWR meter and decay measurement device P2–61. The experimental frequency dependences of $|R|^2$ (●●●●) and $|T|^2$ (▲ ▲ ▲ ▲) for the matched load in the three-centimeter wavelength range are shown in Fig. 4.9. As follows from the results of the experiment, the use of microwave photonic crystals based on metal–dielectric structures with nanometer metal layers to create a matched load makes it possible to provide an $|R|^2$ value of less than 0.5% in the range of 8.4–11.2 GHz, a $|T|^2$ value of less than 0.32% in the 8–12 GHz band. The voltage standing wave ratio (VSWR) in the range 8.4–11.2 GHz was less than 1.05.

The above construction, in which the outer dielectric layer has a small value of the dielectric constant, is protected by a patent [10].

The authors of [11], as a result of solving the optimization problem, including the choice of the surface resistance of a nanometer metal film, have shown the possibility of creating a matched load

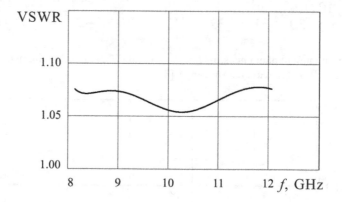

Fig. 4.10. Theoretically calculated frequency dependence of VSWR in the range 8.15–12.05 GHz [11].

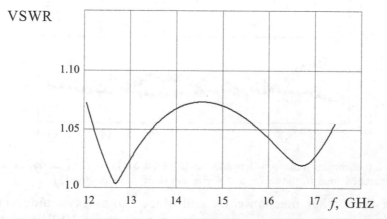

Fig. 4.11. Theoretically calculated frequency dependence of VSWR in the frequency range 12.05–17.44 GHz [11].

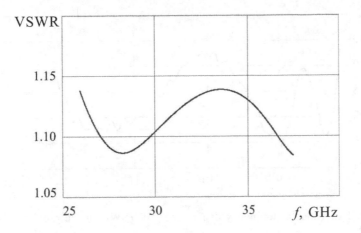

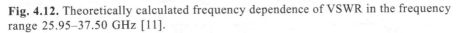

Fig. 4.12. Theoretically calculated frequency dependence of VSWR in the frequency range 25.95–37.50 GHz [11].

of the type described above, providing a magnitude of the volateg standing wave ratio less than 1.10 in the frequency range 8.15–12.05 GHz with a linear load size less than 15 mm (Fig. 4.10). The results of an analytical calculation of the voltage standing wave ratio in the range 12.05–17.44 GHz with its normal incidence on a fully filling waveguide along the cross section of a multilayer metal–dielectric structure consisting of 6 alternating layers are shown in Fig. 4.11. From the results of these calculations, it follows that the use of microwave photonic crystals with nanometer metal layers allows creating a matched load that provides, in the frequency range 12.05–17.44 GHz, the value of the voltage standing wave ratio less than 1.10 for a linear load size of less than 8 mm.

To calculate the frequency dependence of VSWR in the frequency range 25.95–37.50 GHz, a metal–dielectric structure consisting of 7 alternating layers was used. The results of the calculation are shown in Fig. 4.12.

As follows from these results, a VSWR of less than 1.15 can be ensured with a linear load size of less than 6 mm.

According to the above numerical experiment, the matched loads were made for the above-mentioned frequency bands. The design of the load is shown in Fig.4.13.

Nanometric metallic layers (Cr) were deposited on polycarbon substrates. In addition, fluoroplastic layers, layers of expanded Al_2O_3 ceramic and crystalline glass were used. The layers were tightly pressed against each other and placed in a short-circuited section of the waveguide. Measurements of the reflection coefficient and the

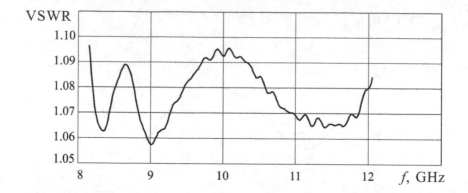

Fig. 4.14, Experimental frequency dependence of VSWR in the range 8.15–12.05 GHz [11].

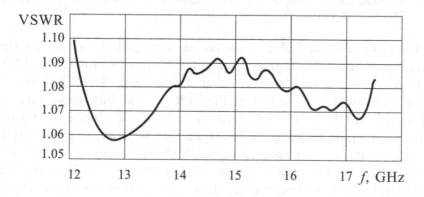

Fig. 4.15. Experimental frequency dependence of VSWR in the range 12.05–17.44 GHz [11].

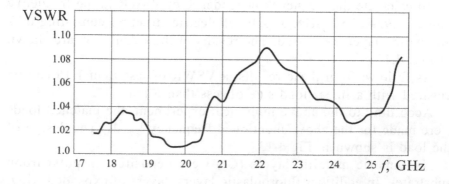

Fig. 4.16. Experimental frequency dependence of VSWR in the range 17.44–25.95 GHz [11].

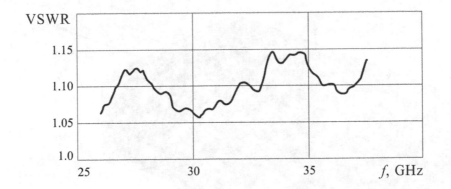

Fig. 4.17. Experimental frequency dependence of VSWR in the range 25.95–37.50 GHz [11].

voltage standing wave ratio were carried out using the Agilent PNA N5230A vector analyzer in the frequency range 8.15–12.05 GHz. The thickness of the metal films was measured on an Agilent AFM 5600LS atomic force microscope. The surface resistance of the metal films was monitored using the Jandel RM3000 probe station.

During the experimental studies in the frequency range 8.15–12.05 GHz (section of the waveguide channel 23×10 mm), 12.05–17.44 GHz (section of the waveguide channel 16×8 mm), 17.44–25.95 GHz (section of the waveguide channel 11×5.5 mm) and 25.95–37.50 GHz (section of the waveguide channel 7.2×3.4 mm), the frequency dependences of the voltage standing wave ratio on the voltage (Fig.4.14–4.17) were measured, interacting with the created matched loads on the basis of multilayer metal–dielectric structures. As follows from the results of the experiment, the use of microwave photonic crystals based on metal–dielectric structures with nanometer metal layers to create a matched load makes it possible to provide a VSWR value of not more than 1.1 in the frequency bands 8.15–12.05 GHz, 12.05–17.44 GHz, 17.44–25.95 GHz. The longitudinal dimensions of the loads were less than 14.5, 10.0, 9.0 mm, respectively. For the frequency range 25.95–37.50 GHz, the VSWR of the realized load was less than 1.15 for a longitudinal dimension of 6.5 mm.

As shown by the results of calculations, in the case of the use of plates from the dielectric materials commonly used in microwave technology, it is by no means always possible to provide the required operating frequency range for sufficiently low VSWR values. In particular, there are difficulties in the selection of materials with a specified with high accuracy value of the dielectric constant from the

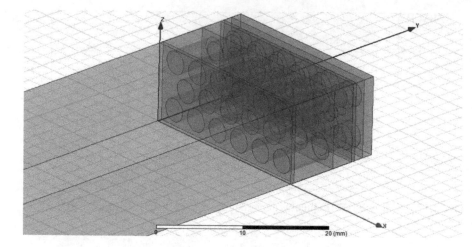

Fig. 4.18. Computer model of the layered structure of the coordinated load located inside the short-circuited section of a rectangular waveguide [10].

range of values 1.1–1.5. In order to avoid these difficulties, a method has been proposed for creating dielectric layers with an effective dielectric constant in a given frequency range, the magnitude of which is determined by the number, arrangement and geometric shape of the air inclusions introduced therein. In the simplest case, such inclusions are through-holes, for example, rectangular or circular [12].

Based on the results of computer modelling, a matched load was created, represented in Figure 4.18, consisting of 7 layers. The first and second layers were made of PTFE, the third, the fourth, the sixth and the seventh – from the ceramic (Al_2O_3), the fifth layer – the nanometer chromium film. The second, third, sixth and seventh dielectric layers contained ordered arrays of air inclusions consisting of 17 through holes of a cylindrical shape, the radius of which was selected during the optimization process to obtain a given value of the effective permittivity. The total length of the layered structure was less than 14 mm.

The described structure made it possible to achieve a VSWR value of less than 1.1 in the frequency bands 8.15–12.05 GHz and 12.05–17.44 GHz. This shows experimentally the possibility of creating broadband matched loads based on layered structures containing one nanometer metal layer and dielectric layers with ordered arrays of air inclusions to change the effective permittivity of these layers, the number and parameters of which are determined by optimization

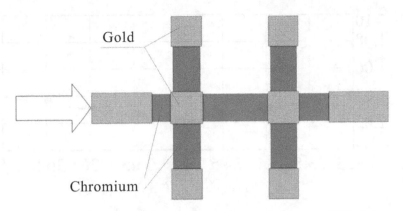

Gold

Chromium

Fig. 4.19. General view of the microstrip matching load [13].

results to obtain the required VSWR value in a given frequency range.

The structure of the microwave photonic crystal can be used to create a small-size broadband matched load in a microstrip version [13]. The authors of [13] proposed the construction of such a load consisting of consecutively connected segments of a microstrip transmission line containing at least seven alternating segments with different surface resistances, the extreme of which are segments with a small surface resistance and at least two pairs of open loops disposed symmetrically on opposite sides of the microstrip line. Each of the loops is made in the form of two consecutively connected segments of a microstrip transmission line with large and small surface resistance (Fig. 4.19). The number of microstrip line segments, their topology and electrical parameters were chosen so that in the selected frequency range the values of the standing wave and transmission coefficients were less than the set values. To calculate the VSWR and the transmission coefficient K_n of the structure, authors [13] carried out computer simulations in the HFSS ANSYS CAD environment.

The results of the calculation are shown in Figs. 4.20 and 4.21, respectively. When solving the optimization problem, the VSWR should be less than 1.1, the transmission coefficient is less than −40 dB. Based on the simulation results, an experimental load sample was produced. The measured frequency dependences of VSWR and K_n are shown in Figs. 4.22 and 4.23, respectively. The measurement results are in good agreement with the simulation results. The authors of [13] showed that by changing the load topology and varying the lengths of segments with large surface resistance of the first and

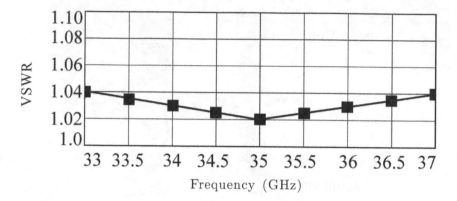

Fig. 4.20. Calculated dependence of the voltage standing wave ratio on frequency in the eight-millimeter wavelength range [13].

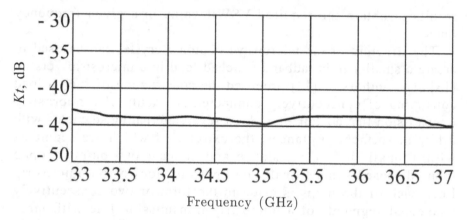

Fig. 4.21. Calculated dependence of the transmittance coefficient on the frequency in the eight-millimeter wavelength range [13]

second pairs of loops, it is possible to achieve even better matching in the chosen frequency range.

It is known to use the spectra of reflection and transmission of electromagnetic radiation to measure the thickness and electrical conductivity of a semiconductor in metal–semiconductor structures [9, 14, 15]. The sensitivity of these methods depends significantly on how much these spectra change when the values of the semiconductor parameters are changed. The determination of these parameters from the results of measurements of spectral dependences is called the solution of the inverse problem. The solution of the inverse problem is simplified if a rigorous theoretical description of the spectral characteristics of the systems under investigation is known. In this case, the inverse problem can be solved as a result of using the

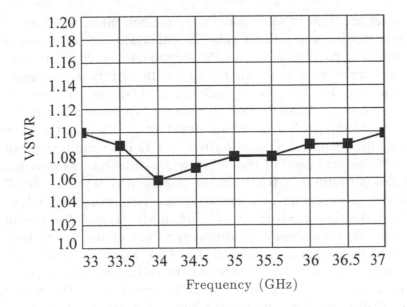

Fig. 4.22. The experimental dependence of the voltage standing wave ratio on frequency in the eight-millimeter wavelength range [13].

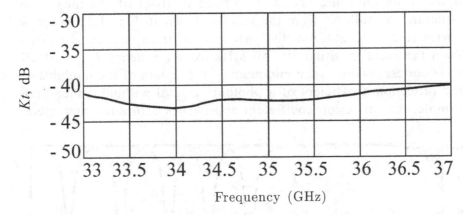

Fig. 4.23. Experimental dependence of the transmission coefficient on the frequency in the eight-millimeter wavelength range [13].

procedure for minimizing the difference between the calculated and experimental dependences. The authors of [16] used measurements of the parameters of nanometer metal layers on insulating substrates to suggest waveguide photonic structures in which the metal–dielectric structure was placed in a segment of a microwave photonic crystal that violated its periodicity. In the 'forbidden' band of a photonic crystal, a 'window' of transparency appears that is sensitive to the desired parameters of the metal–dielectric structure. Such structures are used in microwave acoustoelectronics, in micro- and nanoelectronics and other fields. To calculate the reflection and transmittance coefficients of an electromagnetic wave, transmission matrices were used between regions with different propagation constants [9, 14, 15]. We considered an 11-layer structure of a photonic crystal consisting of alternating layers of a 1 mm thick ceramic (Al_2O_3) with a dielectric constant $\varepsilon_c = 9.6$ and a 12 mm thick plastic foam and a permittivity $\varepsilon_f = 1.1$. To create transparency in the 'forbidden' band of the 'window', a violation was introduced into this structure in the form of one (sixth) layer of smaller thickness. The results of calculating the frequency dependences of the reflection coefficient of a photonic crystal $R(\omega)$ with a 'window' of transparency for different values of the permittivity ε of one of the layers are shown in Fig.4.24. Figure 4.25 shows the calculated frequency dependences of the reflection coefficient of a photonic crystal near the 'window' of transparency in the presence of a violation in it as one (sixth) layer of a smaller thickness d_6 for different values of the thickness of the nanometer metal layer h when the photonic crystal is placed in front of the measured structure. As follows from the results shown in Fig. 4.25, when h varies from 0 to 200 nm, the 'window' of transparency is shifted from the initial position by ~50 MHz. At a frequency $f_{min1} = 10.39$ GHz corresponding to the minimum of the square of the modulus of the reflection coefficient of a photonic crystal without a measured sample, the reflection coefficient approaches saturation at a metal

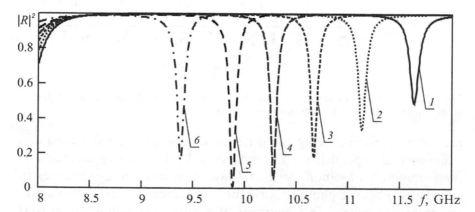

Fig. 4.24. The frequency dependences of the square of the modulus of the reflection coefficient of a photonic crystal for various values of the dielectric constant ε of the 7-th layer at a thickness $d_6 = 5$ mm of the violated 6-th layer (foam): ε, rel. units. *1* – 1, *2* – 3, *3* – 5, *4* – 7, *5* – 9.6, *6* – 15 [16].

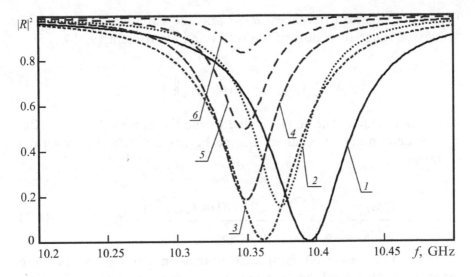

Fig. 4.25. Frequency dependences of the square of the reflection coefficient modulus of a photonic crystal for different values of the thickness h of a nanometer metal layer when the photonic crystal is placed in front of the measured structure: $d_6 = 4$ mm. Curve *1* corresponds to a photonic crystal without a measurable structure. h, nm: $2 - 0$, $3 - 4$, $4 - 20$, $5 - 50$, $6 - 200$ [16].

(chromium) layer thickness of more than 40 nm, while at a frequency $f_{min2} = 10.365$ GHz, corresponding to a minimum of the reflection coefficient of the photonic crystal with the sample being measured, the reflection coefficient approaches saturation at a thickness of the metallic layer (chromium) of more than 150 nm.

Such a resonance reflection character makes it possible to use the frequency dependences of the reflection coefficient of a photonic crystal in the presence of a violation in it as a single layer of a smaller thickness to control the sensitivity of the reflection coefficient to a change in the thickness of the nanometer metal layer of the measured structure.

The use of the frequency dependence of the square of the modulus of the reflection coefficient of electromagnetic radiation $|R(\omega)|^2$ on the photon structure with the sample being measured makes it possible to determine the thickness $t_{m\,req}$ of a nanometer metal film from the solution of equation (1) with a known electrical conductivity.

To determine the required thickness $t_{m\,req}$ of the nanometer metal film from the frequency dependence $|R(\omega)|^2$, the least squares method can be used, in the realization of which the value of the parameter $t_{m\,req}$ is found, at which the sum $S(t_{m\,req})$ of the squares of the experimental differences $|R_{exp}|^2$ and the calculated $|R(\omega,t_{m\ req})|^2$ values of squared moduli of the reflection coefficient [9, 14, 15]

$$S(t_m) = \sum \left(|R_{exp}|^2 - |R(\omega, t_{m\ req})|^2 \right)^2 \tag{4.1}$$

becomes minimal. The sought value of the thickness of a metal film $t_{m\ req}$ is determined by a numerical method as a result of solving the equation

$$\frac{\partial S(t_{m\ req})}{\partial t_{m\ req}} = \frac{\partial \left(\sum \left(|R_{exp}|^2 - |R(\omega, t_{m\ req})|^2 \right) \right)^2}{\partial t_{m\ req}} = 0. \tag{4.2}$$

When the measured sample is placed in front of a photonic crystal, the form of the frequency dependence of the reflection coefficient in the 'window' region of transparency, as follows from the results of the calculations shown in Fig. 4.26, also undergoes significant changes with increasing thickness of the metal layer. At the same time, the sharpest change in the square of the modulus of the reflection coefficient at the minimum of the dependence $|R(\omega)|^2$ is observed in the range of thicknesses of metallic (chromium) films of 0–20 nm.

Measurements were made of $|R(\omega)|^2$ in the frequency range 8–12 GHz for a structure with the parameters specified in the calculation.

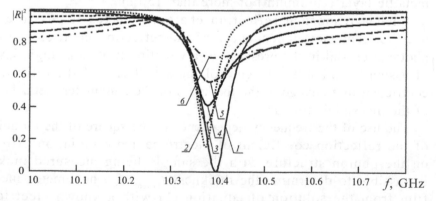

Fig. 4.26. Frequency dependences of the square of the modulus of reflection coefficient of a photonic crystal for various values of the thickness h of a nanometer metal layer when the measured structure is placed in front of a photonic crystal at a thickness d_6=4 mm of the broken 6th layer (polystyrene). Curve *1* corresponds to a photonic crystal without a measured structure. h, nm: *2* – 0, *3* – 2.5, *4* – 5, *5* – 10, *6* – 25 [16].

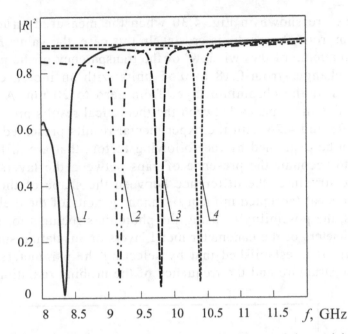

Fig. 4.27. Measured frequency dependences of the square of the modulus of the reflection coefficient of a photonic crystal for various values of the thickness d_6 of the violated 6th layer (a layer of foam): Curve 1 correspond to a photonic crystal without disturbances, d_6, mm: $2 - 7.0$, $3 - 5.0$, $4 - 4.0$. [16].

The measurements were carried out using a panoramic VSWR meter and decay meter P2–61. Figure 4.27 shows the measured frequency dependences of $|R_{exp}|^2$ near the 'window' of transparency for different values of the thickness of the 6th layer (foam).

The results of similar measurements for different values of the permittivity ε of the 7th layer are shown in Fig. 4.28. The presented dependences demonstrate the high sensitivity of the form of the frequency and amplitude dependence $|R(\omega)|^2$ on the dielectric constant of one of the layers of the periodic structure.

The results of measurements of $|R_{exp}(\omega)|^2$ for different values of the thickness of a chromium film deposited on a Al_2O_3 ceramic substrate when the photonic crystal is placed in front of the measured structure, in the presence of a violation in the form of a sixth layer of a smaller thickness d_6, are shown in Fig. 4.29. With an increase in the thickness of the chromium layer from zero to 144 nm, the value $|R_{exp}(\omega)|^2$ at the minimum of the 'window' of the transparency of the photonic crystal varies in the range from 0.03 to 0.5.

The results of measurements of $|R_{exp}(\omega)|^2$ for different values of the thickness of a chromium film deposited on a Al_2O_3 ceramic substrate are shown in Fig. 4.30 when the measured structure is placed in front of a photonic crystal. In this case, the value $|R_{exp}(\omega)|^2$ at the minimum of the 'window' of the transparency of the photonic crystal changes from 0.18 to 0.55 units with an increase in the thickness of the chromium layer from zero to 20 nm. A certain quantitative discrepancy between the theoretical results presented in Figs. 4.25 and 4.26 with the experimental results presented in Fig. 4.28 can be explained by the following factors that are difficult to take into account: the presence of gaps between the layers of the periodic structure, the difference between the shape of the layers from the ideal (accepted in theory), inhomogeneity of the dielectrics.

Thus, the possibility of using the photonic structures to measure the parameters of the nanometer metal layers on insulating substrates is shown. It is established that by selecting the parameters of the photonic structure and the frequency of the probing radiation in the

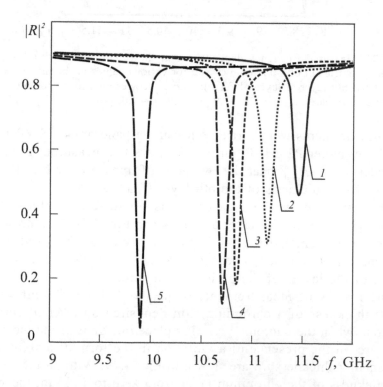

Fig. 4.28. Measured frequency dependences of the square of the modulus of the reflection coefficient of the 11-layer structure of a photonic crystal for various values of the dielectric constant ε of the 7th layer at a thickness of the violated 6th layer $d_6 = 4$ mm. ε, rel. units: *1* – 1 (air gap), *2* – 3 (styrene copolymer ST-3), *3* – 4.3, *4* – 5 (styrene copolymer ST-5), *5* – 9.6 (Al_2O_3 ceramic) [16].

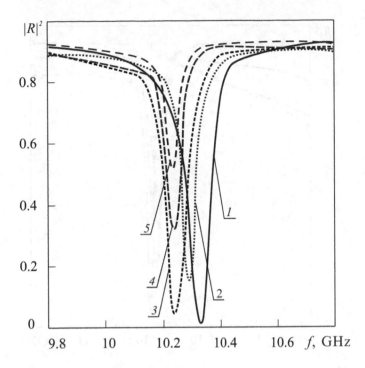

Fig. 4.29. Measured frequency dependences of the squared modulus of the reflection coefficient of a photonic crystal in the presence of a violation in the form of a 6th layer (foam) of a smaller thickness $d_6 = 4$ mm for different values of the thickness h of a nanometer metal layer when the photonic crystal is placed in front of the measured structure: Curve *1* corresponds to a photonic crystal without a measurable structure, h, nm: *2* – 0, *3* – 21, *4* – 76, *5* – 144 [16].

'window' region of the transparency of the photonic structure due to the presence of a disturbance in its periodic structure, it is possible to control the sensitivity of the reflection coefficient to a change in the thickness of the metal layer of the measured structure.

The waveguide microwave methods of measurement using microwave photonic crystals, having high sensitivity and not requiring calibration, allow obtaining a measurement result averaged over a size comparable to the wavelength of the radiation. One of the most successful and promising methods for diagnosing materials and structures that allow measurements with high spatial resolution is near-field microwave microscopy [17–19].

A key element of the near-field microwave microscope is a probe with an aperture size much smaller than the wavelength of the microwave radiation. In near-field microwave microscopes, the field of the non-propagating type of waves is used as a probe [9, 17, 18].

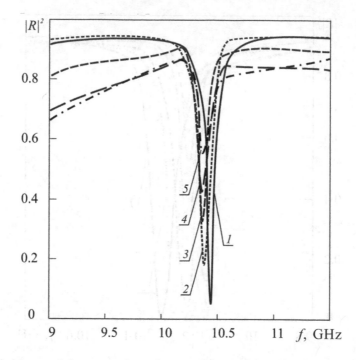

Fig. 4.30. Measured frequency dependences of the modulus of reflection coefficient of a photonic crystal in the presence of a sixth-layer (polystyrene) disruption in it of a lesser thickness ($d_6 = 4$ mm) for different values of the thickness of the nanometer metal layer h when the photonic crystal is behind after the measured structure. Curve *1* corresponds to a photonic crystal without a measured structure, h, nm: *2* – 0; *3* – 12; *4* – 18; *5* – 21. [16]

It is this field that is formed if the central conductor of the coaxial line goes beyond the outer conductor [19]. The advantage of using waves of a non-propagating type is that they attenuate at a small distance, which allows one to obtain a high spatial resolution.

The main element of the near-field microwave microscope, which ensures its high sensitivity and resolving power, is, according to the authors of [17], the microwave resonator connected to the probe.

With increasing sensitivity of the resonator to the perturbation introduced into it through the probe, the sensitivity and resolving power of the microwave microscope as a whole increase.

The possibility of creating microwave resonators based on the so-called low-dimensional resonance systems in which resonances are associated with higher types of oscillations is shown in [20–27] and their increased sensitivity to disturbing effects is detected. The resonator design in conjunction with the probe element for a near-field microwave microscope is described in Ref. [28]. In [29], the results of a study of the possibility of using a one-dimensional photonic crystal in which the load is a tunable resonator that allows the control of resonance singularities in the reflection spectrum of a near-field microwave microscope probe is given. As such a load, the authors of Ref. [29] chose a cylindrical resonator with a communication coupler extending beyond the cavity, the end of which is used as a probe of a near-field microscope to control the parameters of dielectric plates with different values of the permittivity and the thickness of nanometer metal layers deposited on dielectric plates.

A general view of the probe of a near-field microwave microscope on the basis of a cylindrical microwave resonator with a coupling frame element and a one-dimensional photonic crystal is shown in Fig. 4.31. As an element that excites electromagnetic oscillations, a segment of a waveguide *2* with a cross section of 23×10 mm is soldered into the body of the cylindrical resonator *1* so that a membrane *3* with a thickness of 0.5 mm remains between the cavity of the waveguide and the inner wall of the resonator casing. In the membrane *3*, a hole 6 mm in diameter is made through which the frame element *4*, made of copper wire with a diameter of 0.6 mm in cross-section, is passed and intended to connect the cylindrical resonator and waveguide. From the side of the waveguide, the frame is fixed in the foam plate *5*. The frame dimensions are chosen for the optimal transmission of electromagnetic waves in the frequency

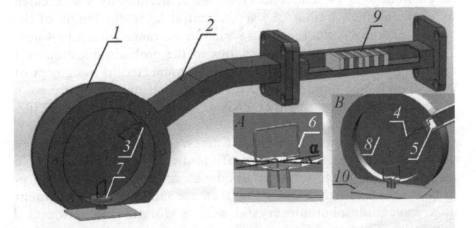

Fig. 4.31. The probe of the near-field microwave microscope on the basis of a cylindrical microwave resonator *1* with a coupling frame *6* and a one-dimensional photonic crystal *9*. Insert A is a frame coupling element. Insert B – a cylindrical microwave cavity with a coupling frame element and a measured sample *10* [28].

range from 8 to 12 GHz [30]. At an angle of 120° to the first frame element *4* with respect to the centre of the circumference of the cylindrical resonator is a second frame element *6* which extends through the aperture in the resonator and whose end part acts as a needle of the probe. The end of the probe was made sharp with a gradually decreasing diameter to a value of 2.0 µm. The second frame element *6* is also made of copper wire with a diameter of 0.6 mm. The element *6* in the cylindrical fluoroplastic sleeve *7* is fixed, which allows to change the position of the frame element by changing the angle α between the plane of the frame and the axis of the probe's needle. The body of the cylindrical resonator (cylinder diameter 65.1 mm, height – 18.3 mm) is closed on both sides with covers *8*.

The magnetic field lines of the field of electromagnetic oscillations in the cavity, crossing the area of the frame coupling element, induce an alternating current in it. Thus, the energy of electromagnetic oscillations is transferred from the volume of the resonator to a small volume in the vicinity of the tip of the probe.

The probe based on a cylindrical resonator with a coupling frame element was connected to a segment of the waveguide photonic crystal 9 with a violation of the periodicity. A one-dimensional waveguide photonic crystal consisting of eleven layers was used in the frequency range 8–12 GHz. Odd layers were made of Al_2O_3 ceramics (Al_2O_3, $\varepsilon = 9.6$), even – from PTFE ($\varepsilon = 2.1$). The length of odd segments – 1 mm, and that of even segments varied in the range from 7 to 14 mm. The violation of periodicity was created by changing the length of the sixth, central layer, the length of the broken sixth layer (fluoroplastic) varied in the range from 3 to 4 mm.

The high-frequency characteristics of the probe in question as a resonator with a communication coupler connected to a segment of a waveguide photonic crystal with a violation of periodicity were investigated using an Agilent PNA-L Network Analyzer N5230A vector analyzer that was connected through a waveguide section.

Figure 4.32 shows the results of measurements of the frequency dependence of the reflection coefficient of the microwave wave in the vicinity of the resonant frequency of the probe in the form of a resonator with a coupling frame connected to a segment of a waveguide photonic crystal with a violation of periodicity (curve *1*). The same figure shows the frequency dependence of the reflection coefficient of the microwave wave in the vicinity of the resonant frequency of the probe in the form of a resonator with a communication coupler without a photonic crystal (curve *2*)

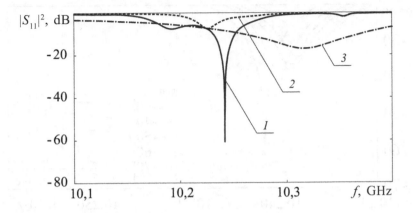

Fig. 4.32. Frequency dependences of the reflection coefficient of the microwave wave in the vicinity of the resonance frequency of the probe in the form of a resonator with a communication coupler connected to a segment of a waveguide photonic crystal with a violation of periodicity (curve *1*), with a communication coupler without a photonic crystal (curve *2*), from of a separately taken waveguide photonic crystal with a violation of periodicity (curve *3*) [28].

and the frequency dependence of the reflection coefficient of the microwave wave of a photonic crystal near its 'transparency window' (curve *3*). As the tip of the probe of the sample is approached, the input impedance of the probe changes abruptly and the reflection coefficient of the microwave wave changes from the probe. The magnitude of its change depends on the parameters of the sample under study, such as: electrical conductivity, dielectric constant, thickness.

By rotating the frame coupling element of the measuring probe, which changes the effective cross-sectional area of the frame penetrated by the magnetic field lines of the electromagnetic oscillation field in the cavity, and which leads to a change in the input impedance of the probe, it is possible to achieve maximum sensitivity of the reflection coefficient of the microwave wave to a change in the electrophysical characteristics of the sample under study. The resonant frequency of the probe changes somewhat.

Figure 4.33 shows the results of measurements of the frequency dependence of the reflection coefficient of the microwave wave in the vicinity of the resonance frequency at a fixed gap (18 μm) between the probe and the samples under study with different permittivity at the angle of rotation of the measuring probe frame α = 178°.

As follows from the presented results, an increase in the relative permittivity of samples from 1 to 11.7 leads to a frequency shift of

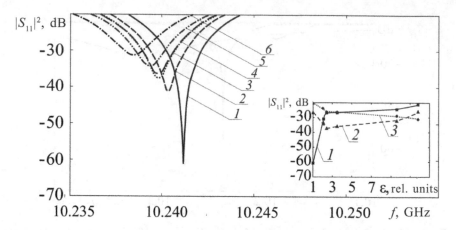

Fig. 4.33. Frequency dependences of the reflection coefficient of the microwave wave in the vicinity of the resonant frequency with a fixed gap equal to 18 μm between the probe and samples with different dielectric permittivity. Curve *1* corresponds to the absence of the measured sample ($\varepsilon = 1$), *2* – fluoroplastic ($\varepsilon = 2.0$), *3* – getinax ($\varepsilon = 2.5$), *4* – textolite ($\varepsilon = 3.4$), *5* – Al$_2$O$_3$ ceramic ($\varepsilon = 9.6$), *6* – silicon ($\varepsilon = 11.7$) [28].

the resonance curve at 3.0 MHz, while the value of the reflection coefficient at the minimum of the resonance curve varies from −60.8 dB to −31.1 dB.

The inset in Fig. 4.33 shows the dependences of the reflection coefficient of the microwave wave, measured at various fixed frequencies in the vicinity of the reflection coefficient minimum, on the dielectric permittivity of the samples placed at a fixed distance near the tip of the probe.

As follows from the presented results, the choice of the frequency of the probing radiation can be obtained either monotonically increasing (curve *1* in the inset in Fig. 4.33) or a monotonically decreasing (curve *3*) dependence of the reflection coefficient of the microwave wave on the dielectric constant of the samples under study, and the non-monotonic curve *2*).

When choosing the frequency at which measurements are taken, corresponding to a minimum of the reflection coefficient in the absence of a measured sample, the range of changes in the reflection coefficient with a change in the permittivity is maximal and is ~39.6 dB. The measured sensitivity $\partial S_{11}/\partial \varepsilon$ monotonically decreases with increasing ε in the range of values from 1 to 11.7. In this case, in the range of values of $\varepsilon = 1 - 2$, $\partial S_{11}/\partial \varepsilon$ is ~29.7 dB/ε, and the resolving power $\Delta\varepsilon/\varepsilon$ reaches a value of ~10⁻⁴.

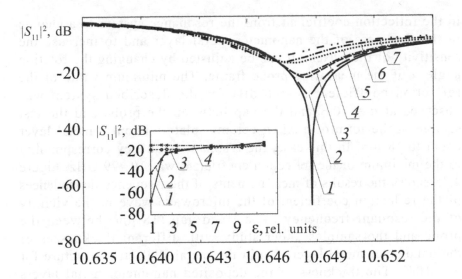

Fig. 4.34. Frequency dependences of the reflection coefficient of the microwave wave in the vicinity of the resonance frequency when the probe is applied to the tip of the test samples with different dielectric permittivity. Curve *1* corresponds to the absence of the measured sample ($\varepsilon = 1$), *2* – fluoroplastic ($\varepsilon = 2.0$), *3* – getinax ($\varepsilon = 2.5$), *4* – textolite ($\varepsilon = 3.4$), *5* – photo glass ($\varepsilon = 6.7$), *6* – Al_2O_3 ceramics ($\varepsilon = 9.6$), *7* – silicon ($\varepsilon = 11.7$) [28].

The results of measurements of the frequency dependences of the reflection coefficient of the microwave wave in the vicinity of the resonance frequency when the samples with different dielectric permittivity are applied to the tip of the probe with the angle of rotation of the measuring probe frame $\alpha = 166°$ are shown in Fig. 4.34.

The inset in Fig. 4.34 shows the dependences of the reflection coefficient of the microwave wave measured at various fixed frequencies (10.64873, 10.64861, 10.64726, 10.64638 GHz) in the vicinity of the reflection coefficient minimum, and the dielectric permittivity of the samples placed at the tip of the probe.

In this case, the measured sensitivity of $\partial S_{11}/\partial \varepsilon$ in the range of $\varepsilon = 1 - 2$ is ~39.5 dB/ε, and the resolving power $\Delta\varepsilon/\varepsilon$ reaches a value of ~10^{-5}.

The established regularities make it possible to realize the possibility of determining the permittivity of samples with high spatial resolution.

The investigated resonance system can also be used to measure samples in the form of dielectric plates with nanometer metal layers of different thicknesses applied. To ensure a monotonous change in the reflection coefficient from the resonance system to a change in the thickness of the nanometer metal layer and to increase the sensitivity of the system, it can be adjusted by changing the rotation angle α of the measuring probe frame. The minimum value of the reflection coefficient (−46.6 dB) for the described system was observed at $\alpha = 104°$ and the gap between the probe and the test sample in the form of a Al_2O_3 ceramic plate without a metal layer equal to 18 μm. In this case, the resonant frequency corresponding to the minimum of the reflection coefficient was 10.779 GHz. Figure 4.35 shows the results of measurements of the frequency dependences of the reflection coefficient of the microwave wave in the vicinity of the resonant frequency for a fixed gap (18 μm) between the probe and the samples under study with different thicknesses of the nanometer metal layer (Cr) in the metal–dielectric structure for $\alpha = 104°$. The thickness of the deposited nanometer metal layers was measured using AFM5600 atomic force microscope (Agilent Technologies).

As follows from the presented results, an increase in the thickness of a nanometer metal layer (Cr) from 0 to 180 nm results in a change in the reflection coefficient at the minimum of the resonance curve from −46.6 dB to −39.1 dB.

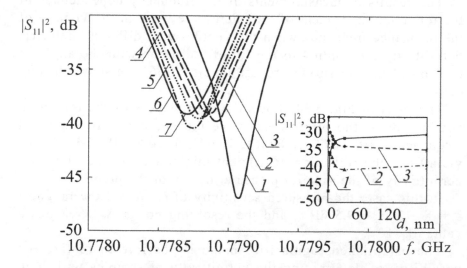

Fig. 4.35. Frequency dependences of the reflection coefficient of the microwave wave in the vicinity of the resonant frequency with a fixed gap equal to 18 μm between the probe and the samples under study with different thicknesses of the nanometer metal layer Cr, nm: *1* – without metallization, *2* – *d* = 3, *3* – *d* = 7, *4* – *d* = 9, *5* – *d* = 13, *6* – *d* = 30, *7* – *d* = 180 [28].

The inset of Fig. 4.35 shows the dependences of the reflection coefficient of the microwave wave measured at various fixed frequencies (10.77908, 10.77895, 10.77850 GHz.) In the vicinity of the reflection coefficient minimum, on the thickness of the nanometer metal layer on Al_2O_3 ceramic plates placed at a fixed distance near the tip of the probe.

The choice of the frequency of the probing radiation can be obtained as a monotonically increasing (curve *1* in the inset in Fig. 4.35) or a monotonically decreasing (curve *3*) dependence of the reflection coefficient of the microwave wave on the thickness of the nanometer metal layer deposited on the Al_2O_3 ceramic plate, and the non-monotonic dependence (curve *2*). The maximum range of the reflection coefficient change was ~16.5 dB. The measured sensitivity $\partial S_{11}/\partial d$ decreased monotonically with increasing thickness of the metallic layer (Cr) *d* in the range 0–180 nm. In the range *d* 0–3 nm, the value of $\partial S_{11}/\partial d$ was ~4.0 dB/nm, and the resolving power reached values of ~10^{-3}. When measuring TaAlN films deposited on a Al_2O_3 ceramic substrate, the sensitivity $\partial S_{11}/\partial d$ was 1.35 dB/nm in the range of values of *d* = 0–20 nm. Figure 4.36 shows the results of measurements of the frequency dependences of the reflection coefficient of the microwave wave in the vicinity of the resonant frequency when the samples with different thicknesses of

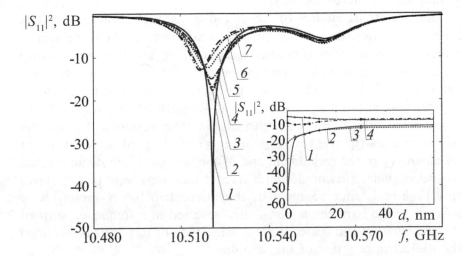

Fig. 4.36. Frequency dependences of the reflection coefficient of the microwave wave in the vicinity of the resonant frequency when the probe is applied to the tip of the test samples with different thicknesses of the nanometer metal layer Cr, nm: *1* – without metallization, *2* – *d* = 3, *3* – *d* = 7, *4* – *d* = 9, *5* – *d* = 13, *6* – *d* = 30, *7* – *d* = 180 [28].

the nanometer metal layer (Cr) are applied to the tip of the probe at a rotation angle of the measuring probe frame of $\alpha = 107°$. In this case, $\partial S_{11}/\partial \varepsilon$ in the range of $d = 0$–3 nm was ~10.6 dB/nm, and the resolving power $\Delta d/d$ reached ~10^{-3}. That is, the proposed method can realize the ability to control the thickness of nanometer metal layers both at a fixed distance of the structure near the tip of the probe and in the mode of touching the metal surface by the probe.

When implementing microwave methods for measuring the parameters of materials, in particular, the materials of the substrates of microwave circuits, microstrip lines are widely used. Their use makes it possible to combine the high sensitivity of microwave measurement methods with the manufacturability of the structures and mandrels for measurement, and the absence of rigid requirements for the dimensions of the samples. The disadvantages of measuring systems on the open microwave transmission lines include the presence of radiation losses at the open ends of the lines, inhomogeneities, for example, related to the need to use coaxial-microstrip transitions, difficulties in carrying out local measurements. The use of microstrip photonic crystals to determine the parameters seems promising in connection with the high sensitivity of the frequency dependence of the 'windows' of transparency that are specially formed in them in the forbidden band to the violation parameters of the periodicity.

The results of studies of the one-dimensional photonic crystals based on a microstrip structure with a substrate of Al_2O_3 ceramics, which consist in series-connected alternating segments of a microstrip transmission line with a varying width of a strip conductor, are presented in [31, 32]. The violation of the periodicity of the photonic crystal was created by decreasing the length of one of the high-resistance (with a smaller width of the strip conductor) segments of the microstrip transmission line, which caused a 'window' of transparency in the forbidden band of the microstrip photonic crystal. Measured dielectric samples with fixed dimensions were placed above this high-resistance segment of the microstrip line symmetrically relative to the strip conductor, which resulted in a frequency shift of the 'window' of transparency. The magnitude of this shift determined the dielectric constant of the sample.

To determine the electrophysical parameters of both nonpolar and polar liquids possessing significant values of the real and imaginary parts of the permittivity in the microwave range, the authors of [33, 34] proposed to use as a heterogeneity of the microstrip photonic structure the segment of the microstrip air transmission line included in it, filled with the test liquid. Even segments of the used photonic crystal were realized on a substrate of Al_2O_3 ceramics, and odd – in the form of segments of strip line with air filling, in which there was an air gap between the strip and the metal base (see Fig.2.13).

The transmission coefficient was calculated usin a transmission matrix [35].

The frequency dependence of the transmission coefficient D of the structure in the frequency range 0–20 GHz calculated in the quasistatic approximation is shown in Fig. 4.37 (curve *1*), and the frequency dependence D is shown in this figure without disrupting the periodicity (curve *2*). The frequency dependences are characterized by the presence of regions forbidden for the propagation of an electromagnetic wave – forbidden bands in microwave photonic crystals. On the frequency dependence of the transmittance, the location of the 'windows' of transparency, the appearance of which is caused by the presence of an inhomogeneity in the microstrip structure, can be controlled by changing the parameters of this inhomogeneity, for example, by placing a cuvette with a liquid with different dielectric permittivity (Fig. 4.38).

For the experimental studies we used structures with lengths of the 1st, 3rd, 7th and 9th segments of 20 mm, the 2nd and the 8th – 8 mm, the width of the strip conductor – 1 mm, the thickness of the dielectric substrate – 1 mm. A wide cuvette was used to measure liquids with low losses (Fig. 4.39 *a*) with the dimensions of the 4th, 5th and 6th segments of 15.5 mm, 0.8 mm and 15.5 mm, respectively. For liquids

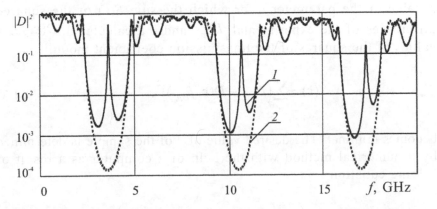

Fig. 4.37. Calculated frequency dependences of the transmittance of the microstrip photonic structure with the violation of periodicity in the form of the changed lengths of the 4th, 5th and 6th segments (curve *1*) and without violation of periodicity (curve *2*) [33].

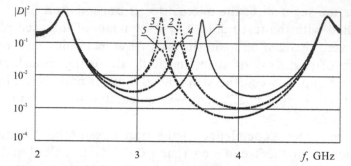

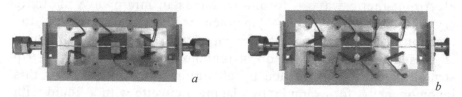

Fig. 4.38. The calculated frequency dependences of the transmittance D of the microstrip photonic structure with a cuvette filled with substances with different values of the permittivity (ε, rel. units): $1 - 1.0$ (empty cuvette), $2 - 40$, $3 - 80$, $4 - 40-5j$, $5 - 80-10j$ [33].

Fig. 4.39. Experimental microband photonic structure. a) with a wide cuvette; b) with a narrow cuvette [33].

with high losses, a narrow cuvette (Fig. 4.39 *b*) with dimensions of the 4th, 5th and 6th segments of 14 mm, 10 mm and 8 mm.

To determine the complex dielectric constant $\varepsilon = \varepsilon' - j \cdot \varepsilon''$ of the liquid filling the cuvette, the inverse problem was solved using the method of least squares. When implementing this method, there was a value of the parameter ε for which the sum $S(\varepsilon)$ of the squared differences of the experimental D_{exp_n} and the calculated $D(\varepsilon, f_{exp_n})$ values of the squares of the transmission coefficient moduli

$$S(\varepsilon) = \sum_n \left(D_{exp_n} - D(\varepsilon, f_{exp_n}) \right)^2, \tag{4.3}$$

becomes minimal. The desired value of ε of the sample is determined by a numerical method with the help of a computer as a result of solving equation

$$\frac{\partial S(\varepsilon)}{\partial \varepsilon} = \frac{\partial \left(S(\varepsilon) = \sum_n \left(D_{exp_n} - D(\varepsilon, f_{exp_n}) \right)^2 \right)}{\partial \varepsilon} = 0. \tag{4.4}$$

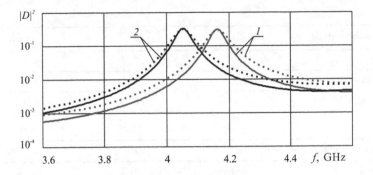

Fig. 4.40. Frequency dependences of transmission coefficients of a microstrip photonic structure with an inhomogeneity (see Fig. 4.39 *a*) in the form of an empty cuvette (curves *1*) and a cuvette filled with a nonpolar liquid (transformer oil) (curves *2*): measured – dotted curves, calculated – continuous curves [33].

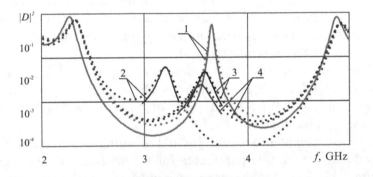

Fig. 4.41. Frequency dependences of transmission coefficients of a microstrip photonic structure with inhomogeneity (Fig. 4.39 *b*) in the form of an empty cuvette (curves *1*) and a cuvette filled with polar liquids: deionized water (curves *2*), ethyl alcohol (curves *3*), glycerol curves *4*): measured – dashed curves, calculated – continuous curves [33].

Figure 4.40 shows the measured (dashed) frequency dependences D for a structure with an inhomogeneity (see Fig. 4.39 *a*) as an empty cuvette and a cuvette filled with a nonpolar liquid (transformer oil).

Figure 4.41 presents analogous dependencies for a structure with an inhomogeneity (see Fig.4.39, *b*) as an empty cell (curves *1*) and a cuvette filled with polar liquids: deionized water (curves *2*), ethyl alcohol (curves *3*), glycerol (curves *4*). The measured frequency dependences D near the 'window' of transparency were used to determine the ε' and ε'' values of the liquid dielectrics from the solution of equation (4.4).

Figures 4.40 and 4.41 also show the calculated frequency dependences $|D(\varepsilon, f_{\exp_n})|^2$ (continuous curves) for the values of

Table 1

Non-polar liquids	ε'	ε''	f_{max}, GHz
Toluene	2.43	0.025	4.06
Transformer oil	2.5	0.026	4.05
Vacuum oil BM-5	2.4	0.017	4.06

Table 2

Polar fluids	ε'	ε''	f_{max}, GHz
Deionized water	76.6	10.3	3.24
Ethanol (96% vol.)	10.5	7.9	3.58
Glycerol	16	11.5	3.54

ε given in the Tables 1 and 2, which were determined from the solution of equation (4.4) at frequencies f_{max} corresponding to the maximum transmittance coefficient $D(\varepsilon, f_{exp_n})$ in the 'window' of the transparency of the photonic crystal.

It should be noted that since the polar liquids exhibit a strong frequency dependence $\varepsilon(f)$, calculations of the frequency dependence of the transmittance for a fixed value of ε are correct only near the frequency f_{max} (see Fig. 4.41).

The measurement procedure for determining ε described above was used to measure this parameter for water–ethanol solutions.

Figure 4.42 shows the frequency dependences D measured near the 'window' of transparency for the structure with a cuvette (see Fig. 4.39 b) filled with a water–ethanol solution with different volume content X of the ethanol.

The measured frequency dependences were used to determine the value of ε of the solutions by solving the equation (4.4).

Figure 4.42 also shows the calculated frequency dependences $|D(\varepsilon, f_{exp_n})|^2$ (solid curves) for the values of ε determined from the solution of Eq. (4.4) for solutions with different volume content of ethanol at frequencies f_{max} corresponding to the maximum transmission coefficient $D(\varepsilon, f_{exp_n})$ in the 'window' of the transparency of the microstrip photonic crystal.

The results of measurements of ε' and ε'' of the solutions with different volume content of ethanol are shown in Fig. 4.43.

The obtained measurement results ε' and ε'' are in good agreement with the results of the developed model for describing the spectra of ε such solutions, taking into account the change in the polarization

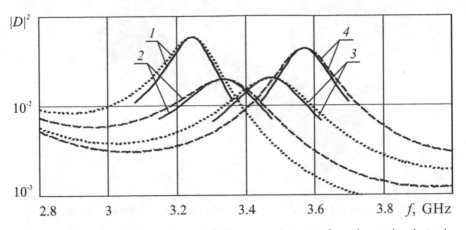

Fig. 4.42. Frequency dependences of the transmittance of a microstrip photonic structure with an inhomogeneity (see Fig, 4.39 *b*) in the form of a cuvette filled with a water–ethanol solution with different volume content of ethanol, *X*: *1* – 0; *2* – 0.38; *3* – 0.67; *4* – 0.96. Measured – dashed curves, calculated – continuous curves [33].

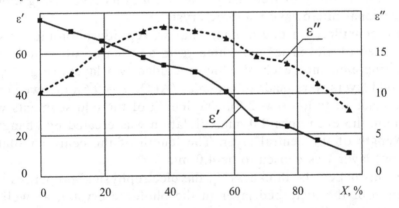

Fig. 4.43. Experimental dependences of ε' and ε'' of the water–ethanol solution on the volume fraction *X* of the ethanol in the solution [33].

relaxation time of the water-ethanol solution with the change in the volume fraction of ethanol [36].

One of the most important stages in the process of creating modern devices for micro- and nanoelectronics is the implementation of multi-parameter control of the electrophysical characteristics and layer thicknesses of the real semiconductor structures. Of particular interest is the simultaneous measurement of the thickness of the substrate of a semiconductor structure, the thickness and specific conductivity of a heavily doped epitaxial layer and the mobility of free charge carriers in this layer.

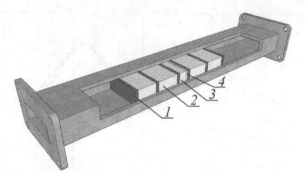

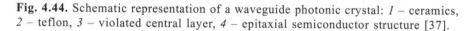

Fig. 4.44. Schematic representation of a waveguide photonic crystal: *1* – ceramics, *2* – teflon, *3* – violated central layer, *4* – epitaxial semiconductor structure [37].

In [37], it is proposed to implement a method for simultaneous measurement of the substrate thickness of a semiconductor structure, the thickness and specific conductivity of a heavily doped epitaxial layer, the mobility of free charge carriers in this layer using a one-dimensional microwave photonic crystal.

We considered a one-dimensional waveguide photonic crystal composed of eleven layers, which form a structure of periodically repeating elements, each of which included two layers (Fig. 4.44). The odd layers were made of ceramics (Al_2O_3, $\varepsilon = 9.6$), the even ones were made of teflon ($\varepsilon = 2.0$). The length of the odd segments was 1.0 mm, the even ones 9.0 mm. Violation was created by changing the length of the central layer. The length of the central violated (teflon) layer was chosen to be 4.0 mm.

In this paper, the thickness d_1, the electrophysical properties and position of the disturbed layer in the photonic crystal, as well as the position of the measured structure inside the disturbed layer, were fixed. The measured structure placed on the boundary of the disturbed central layer of the fluoroplastic and the next layer of Al_2O_3 ceramics was oriented in two ways with respect to the direction of propagation of the electromagnetic wave. The location of the sample inside the disturbed layer and its orientation relative to the disturbed layer in the photonic crystal are shown in Fig. 4.45 (configurations 1 and 2).

The parameters of a photonic crystal with a violation of periodicity were selected in such a way that the forbidden band covered a large part of the 3-cm range of wavelengths, and the transmission peak was located in the middle of the forbidden band.

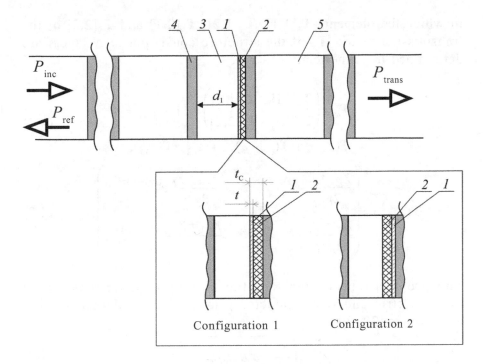

Fig. 4.45. Location of the semiconductor structure with respect to the disturbed layer in the waveguide microwave photonic crystal: *1* – heavily doped semiconductor layer, *2* – high-resistance substrate, *3* – violated central layer, *4* and *5* – periodically alternating layers with different values of the dielectric constant [37].

The investigated samples consisted of epitaxial arsenide gallium structures of thickness $t_s = t + t_{sub}$, consisting of an epitaxial layer of thickness t with electrical conductivity σ and a semi-insulating substrate of thickness t_{sub}.

The frequency dependence of the reflection $R(\omega)$ and transmission coefficients $D(\omega)$ of an electromagnetic wave with its normal incidence on a multilayer structure completely filling the waveguide along the cross section and having plane layers perpendicular to the propagation direction of radiation, is calculated using the expressions

$$R = \frac{\mathbf{T}_N[2,1]}{\mathbf{T}_N[2,2]}, \tag{4.5}$$

$$D = \frac{\mathbf{T}_N[1,1] \cdot \mathbf{T}_N[2,2] - \mathbf{T}_N[1,2] \cdot \mathbf{T}_N[2,1]}{\mathbf{T}_N[2,2]}, \tag{4.6}$$

in which the elements $\mathbf{T}_N[1,1]$, $\mathbf{T}_N[1,2]$, $\mathbf{T}_N[2,1]$ and $\mathbf{T}_N[2,2]$ of the transmission matrix $\mathbf{T}_N$ of the structure consisting of N layers are determined from the relation

$$\mathbf{T}_N = \begin{pmatrix} \mathbf{T}_N[1,1] & \mathbf{T}_N[1,2] \\ \mathbf{T}_N[2,1] & \mathbf{T}_N[2,2] \end{pmatrix} = \prod_{j=N}^{0} \mathbf{T}_{j,(j+1)} =$$

$$\mathbf{T}(z_{N,N+1}) \cdot \mathbf{T}(z_{N-1,N}) \ldots \mathbf{T}(z_{1,2}) \cdot \mathbf{T}(z_{0,1}), \tag{4.7}$$

$$\mathbf{T}(z_{j,j+1}) = \begin{pmatrix} \dfrac{\gamma_{j+1} + \gamma_j}{2\gamma_{j+1}} e^{(\gamma_{j+1} - \gamma_j)z_{j,j+1}} & \dfrac{\gamma_{j+1} - \gamma_j}{2\gamma_{j+1}} e^{(\gamma_{j+1} + \gamma_j)z_{j,j+1}} \\ \dfrac{\gamma_{j+1} - \gamma_j}{2\gamma_{j+1}} e^{-(\gamma_{j+1} + \gamma_j)z_{j,j+1}} & \dfrac{\gamma_{j+1} + \gamma_j}{2\gamma_{j+1}} e^{-(\gamma_{j+1} - \gamma_j)z_{j,j+1}} \end{pmatrix}. \tag{4.8}$$

The propagation constants of the electromagnetic wave γ_d in the dielectric and semiconductor γ_s layers were calculated using expressions [38, 39]:

$$\gamma_d = \sqrt{\frac{\pi^2}{a^2} - \omega^2 \varepsilon_d \varepsilon_0 \mu_0}, \tag{4.9}$$

$$_s = \sqrt{\frac{}{} - \omega^2 \varepsilon_s^* \varepsilon_0 \mu_0} \tag{4.10}$$

where $\varepsilon_s^* = \varepsilon_s' - j\varepsilon_s''$ is the complex dielectric constant of the semiconductor layer; $\varepsilon_s' = \varepsilon_s - \dfrac{\sigma_s^2 m_s^*}{\varepsilon_0 e^2 n_s}$, $\varepsilon_s'' = \dfrac{\sigma_s}{\varepsilon_0 \omega}$ are real and imaginary parts of the complex dielectric permittivity of the semiconductor layer; ε_s is the relative dielectric permittivity of the semiconductor layer; σ_s is the electrical conductivity of the semiconductor layer; m_D^* and n_s are the effective mass and concentration of electrons in the semiconductor layer; a is the size of the wide wall of the waveguide; $\omega = 2\pi f$ is the circular frequency of the electromagnetic wave; ε_0 and μ_0 are the dielectric and magnetic permeabilities of vacuum; ε_d is the relative permittivity of the dielectric layer.

For the simultaneous determination of the thickness of the semi-insulating substrate t_{sub}, the thickness t, and the electrical conductivity σ of the conductive layer the least squares method was used for the frequency dependences $D(\omega)$ and $R(\omega)$ to determine the values of the parameters t_{sub}, t, and σ for which the sum $S(t_{\text{sub}}, t, \sigma)$ of the squared difference of the calculated, $|D(\omega,t_{sub},t,\sigma)|^2$ and $|R(\omega,t_{sub},t,\sigma)|^2$, and the experimental, $|D_{exp}|^2$ and $|R_{exp}|^2$, transmission and reflection coefficients, measured with two different configurations (Fig.4.45) is violated the photonic crystal

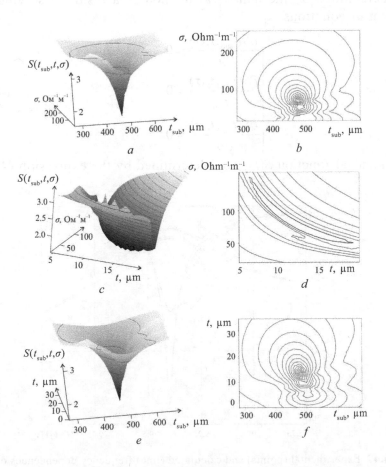

Fig. 4.46. The form of the residual function in space and contour maps in the planes of the required parameters: a, $b - (t_{sub},\sigma)$; c, $d - (t, \sigma)$; e, $f - (t_{sub}, t)$ for a sample of gallium arsenide structure with an epitaxial layer of $t = 13.14$ μm and electrical conductivity $\sigma = 71.73$ Ohm^{-1}m^{-1} grown on a high-resistance substrate with a thickness $t_{sub} = 480.3$ μm [37].

$$S(t_{sub},t,\sigma) = \sum_{i=1}^{K} \begin{pmatrix} \left(|D_1(\omega_i,t_{sub},t,\sigma)|^2 - |D_{i1exp}|^2\right)^2 + \\ +\left(|R_1(\omega_i,t_{sub},t,\sigma)|^2 - |R_{i1exp}|^2\right)^2 + \\ +\left(|D_2(\omega_i,t_{sub},t,\sigma)|^2 - |D_{i2exp}|^2\right)^2 + \\ +\left(|R_2(\omega_i,t_{sub},t,\sigma)|^2 - |R_{i2exp}|^2\right)^2 \end{pmatrix} \quad (4.11)$$

becomes minimal. Here K is the number of measured values of the transmission and reflection coefficients.

The required values of the parameters of the investigated sample are determined by the numerical method as a result of solving the system of equations:

$$\begin{cases} \dfrac{\partial S(t_{sub},t,\sigma)}{\partial t} = 0 \\[2ex] \dfrac{\partial S(t_{sub},t,\sigma)}{\partial t_{sub}} = 0 \\[2ex] \dfrac{\partial S(t_{sub},t,\sigma)}{\partial \sigma} = 0 \end{cases} \quad (4.12)$$

The residual function $S(t_{sub},t,\sigma)$, determined by the expression (4.11),

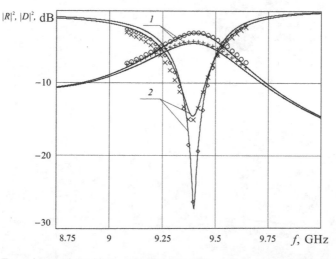

Fig. 4.47. Experimental (points) and calculated (line) frequency dependences of $|D|^2$ (curves *1*) and $|R|^2$ (curves *2*) for two configurations of the photon structure (see the inset to Fig. 4.45: configuration 1 – ◊◊◊ and ○○○; configuration 2 – xxx and +++) containing a sample of a gallium arsenide structure with an epitaxial layer of thickness $t = 13.14$ μm and electrical conductivity $\sigma = 71.73$ Ohm^{-1}m^{-1} grown on a high-resistance substrate with a thickness of $t_{sub} = 480.3$ μm [37].

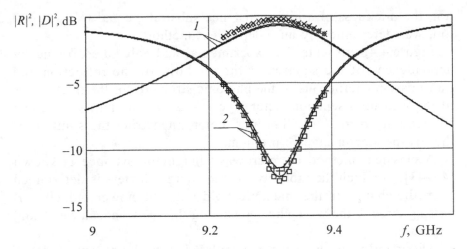

Fig. 4.48. The experimental (points) and calculated (line) frequency dependences of $|D|^2$ (curves *1*) and $|R|^2$ (curves *2*) for the two configurations (see the inset in Fig. 4.46, configuration 1 – ◊◊◊ and □□□; configuration 2 – xxx and +++) of a photonic structure containing a sample of a gallium arsenide structure with an epitaxial layer of thickness $t = 2.17$ μm and electrical conductivity $\sigma = 34.78$ Ohm^{-1}m^{-1} grown on a high-resistance substrate with a thickness $t_{sub} = 480.3$ μm [37].

for the case when the thickness t of the heavily doped GaAs layer was 13.14 μm, its electrical conductivity $\sigma = 71.73$ Ohm^{-1}m^{-1}, and the substrate thickness $t_{sub} = 480.3$ μm, has a global minimum in the coordinate space $(t_{sub}, t, \sigma, S(t,\sigma))$, and the contour maps (Fig. 4.46) are characterized by the presence of closed trajectories near the minimum, which confirms the possibility of uniquely determining the thickness and electrical conductivity of the semiconductor layer from the solution of the system of equations (4.12).

Figure 4.47 shows the experimental and calculated frequency dependences of the moduli of the reflection and transmission coefficients for two configurations of the photonic structure (see the inset in Fig. 4.45) at values of the thickness of the high-resistivity substrate $t_{sub} = 480.3$ μm, the thickness of the semiconductor layer $t = 13.14$ μm and its electrical conductivity $\sigma = 71.73$ Ohm^{-1}m^{-1}, determined from the solution of the inverse problem using the system of equations (4.12).

The obtained thicknesses of the substrate and the epitaxial layer of gallium arsenide are well correlated with the values measured by independent methods.

In the case of small thicknesses of a high-alloy semiconductor layer, the error in determining three parameters is simultaneously significant, therefore, for an epitaxial layer with a thickness $t = 2.17$

μm, the thickness of a high-resistance substrate, equal to $t_{sub} = 482$ μm, was determined by an independent method.

Figure.4.48 presents the experimental and calculated frequency dependences of the squares of the modulus of the reflection and transmission coefficients of the photonic structure for the thicknesses of the measured semiconductor layer $t = 2.17$ μm and its electrical conductivity $\sigma = 34.78$ Ohm^{-1}m^{-1}, determined from the solution of the two-parameter inverse problem.

A classical method of microwave magnetoresistance is known [40–43], in which the mobility of free charge carriers is determined from the change in the magnetic field with the power induction B passing through the waveguide containing the semiconductor epitaxial structure.

The mobility in this case is calculated from the measured decay values α_m and α of the microwave signal in the waveguide section containing a semiconductor structure completely filling the waveguide along a narrow wall and located at the centre of the wide wall of the waveguide, in the presence of an external magnetic field, the magnetic induction vector $\vec{B}$ of which is directed perpendicular to the narrow wall of the waveguide, and without it, respectively:

$$\mu_n = \frac{1}{B}\sqrt{\frac{\alpha - \alpha_m}{\alpha_m}}. \tag{4.13}$$

In this classical method, it is assumed that the transverse dimensions of the epitaxial structure are so small that reflection from the end plate can be neglected.

Using this classical technique, the mobility of the free charge carriers in heavily doped epitaxial layers of gallium arsenide structures was measured at various frequencies of the 3-cm range (Fig. 4.49).

It can be seen that the results of the free carrier mobility measurements performed at different frequencies in the 8–12 GHz band vary greatly: for a semiconductor layer with a thickness $t = 2.17$ μm, the free carrier mobility varied from 0.40 to 0.75 m^2/(V·s), for a semiconductor layer with a thickness $t = 13.14$ μm – from 0.63 to 0.705 m^2/(V·s).

Such a noticeable difference in the result obtained from the frequency at which measurements of the decay of the microwave signal in the waveguide section containing the structure under study

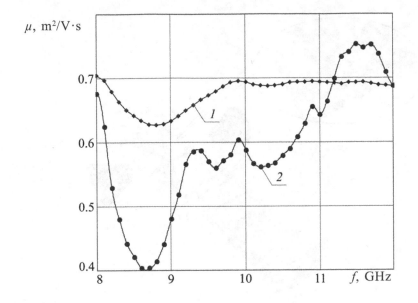

Fig. 4.49. The results of measurements of the mobility of free charge carriers at various frequencies. The thickness of a heavily doped epitaxial layer t, µ0m: *1 –* 13.14, *2 –* 2.17. [37].

were made is associated with the neglect of reflection from the front end of the plate.

It should be noted that the results of measurements of the mobility of charge carriers by the microwave magnetoresistance method at small thicknesses of a heavily doped layer are affected by the size resonance caused by the finite length of the semiconductor structure and unaccounted for in this method.

Thus, as follows from the results presented in Fig.4.49, the relative error in determining the mobility, calculated only from the measured attenuation values α_m and α of the microwave signal in the waveguide section using expression (4.13), was ±22.7% for the gallium arsenide layer with thickness $t = 2.17$ µm and ±4.49% for a layer with thickness $t = 13.14$ µm. We note that a slight decrease in the relative measurement error for the heavily doped layer with a larger thickness is associated with a weakening of the influence that is not taken into account in this method of dimensional resonance due to a significant damping of the wave over the length of the epitaxial structure.

The problem of measuring the mobility of the free charge carriers by the microwave magnetoresistance method can be solved more rigorously if reflection from the front end of the plate is taken

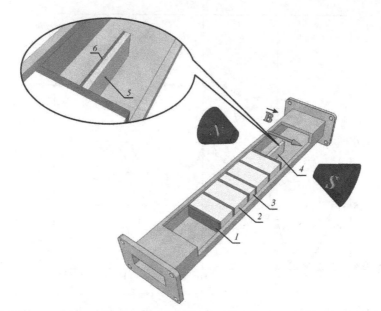

Fig. 4.50. Location of the photonic crystal and epitaxial semiconductor structure in the waveguide: *1* – Al$_2$O$_3$ ceramic layer, *2* – fluoroplastic layer, *3* – violated fluoroplastic layer, *4* – measured gallium arsenide structure, which includes: *5* – high-resistance substrate, *6* – heavily doped semiconductor layer. *N* and *S* are poles of the electromagnet [37].

into account. However, in this case, to calculate the reflection and transmission coefficients, it is necessary to know the parameters of the semiconductor structure, such as the electrical conductivity and the thickness of the layers.

Measurements of the carrier mobility were carried out in epitaxial layers of gallium arsenide structures, the thickness of the layers and electrical conductivity of which were measured using the technique described above with the arrangement of the measured structure inside the violated layer of the microwave photonic crystal (see Fig. 4.45).

To determine the mobility of free charge carriers μ of the semiconductor heavily doped layers, the epitaxial semiconductor structure studied was extracted from the violated central layer of the waveguide photonic crystal and placed in the *E*-plane at the centre of the cross section of the rectangular waveguide after the waveguide photonic crystal. The magnetic induction vector $\vec{B}$ of the magnetic field was directed perpendicular to the narrow walls of the waveguide (Fig.4.50).

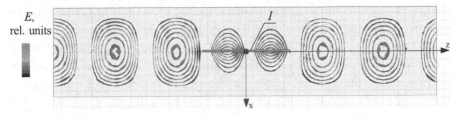

Fig. 4.51. Distribution of electric field strength E of an electromagnetic wave in a segment of a waveguide containing a longitudinally arranged semiconductor structure (*1*) [37].

To describe the interaction of an electromagnetic wave with the structure shown in Fig.4.50, it is possible to use a transmission matrix whose elements are determined through the propagation constants of the electromagnetic wave at each of the sections of the waveguide structure.

Figure 4.51 shows the distribution of the electric field of an electromagnetic wave in a segment of a waveguide containing a semiconductor structure that completely fills the waveguide along a narrow wall and is located at the center of a wide waveguide wall, modelled by the HFSS (High Frequency Structure Simulator).

As follows from the results of the calculation presented in Fig. 4.51, the distribution of the electric field strength undergoes changes in the region with the longitudinal structure of the semiconductor layer, mainly related to the increase of the effective permittivity of this section.

Such a transformation of the field makes it possible to use for the description of the interaction of an electromagnetic wave in this region a wave of the basic type with an effective propagation constant, which can be calculated for a sufficiently fine structure, for example, by the perturbation theory [38].

If the parameters of the epitaxial structure satisfy the relations: $t \ll t_s$ and $\sigma \gg \sigma_s$, the thickness t_s of the measured structure is much smaller than the size a of the wide waveguide wall, then the propagation constant of the wave in the waveguide section containing the longitudinally located semiconductor structure under the action of an external magnetic field can be represented in the form [37, 44]:

$$\gamma_m = \frac{k_0^2 \sigma t}{\beta_0 \omega \varepsilon_0 a} \cdot \frac{1}{1 + \mu^2 B^2} + i\left(\beta_0 - \frac{k_0^2 \sigma t}{\beta_0 \varepsilon_0 a} \cdot \frac{\tau(1 - \mu^2 B^2)}{(1 + \mu^2 B^2)^2} + \frac{k_0^2(\varepsilon_s - 1)t_s}{\beta_0 a} \right) \quad (4.14)$$

where k_0 and β_0 are the phase constants of propagation of an electromagnetic wave in free space and in an empty waveguide, respectively, ε_0 is the dielectric constant of vacuum, $\omega = 2\pi f$ is the circular frequency of the electromagnetic wave, a is the size of the wide wall of the waveguide, ε_s is the relative dielectric constant of the lattice of a semiconductor layer, σ is the electrical conductivity of a semiconductor layer, t_s is thickness of the measured structure, t is the thickness of the semiconductor layer; τ is the relaxation time of the momentum of the free charge carriers.

When describing the microwave photonic crystal and the semiconductor structure located longitudinally behind it, an additional factor appears in the transmission matrix of the layered structure of the form (4.7) in the form of a transmission matrix between the region with the effective propagation constant described by expression (4.14) and the region of the microwave photonic crystal.

The mobility of free charge carriers with respect to the frequency dependences $D(\omega)$ and $R(\omega)$ was determined by the least squares method, in the realization of which a mobility value μ corresponding to the minimum value of the sum $S(\mu)$ of the squares of the difference $|D(\omega,\mu)|^2$ and $|R(\omega,\mu)|^2$ and experimental $|D_{exp}|^2$ and $|R_{exp}|^2$ values of the squares of the moduli of the transmission and reflection coefficients, measured with and without a magnetic field is determined

$$S(\mu) = \sum_{i=1}^{N} \left(\begin{aligned} & \left(|D(\omega_i,\mu)|^2 - |D_{iexp}|^2 \right)^2 + \\ & + \left(|D_B(\omega_i,\mu)|^2 - |D_{iB\,exp}|^2 \right)^2 + \\ & + \left(|R(\omega_i,\mu)|^2 - |R_{iexp}|^2 \right)^2 + \\ & + \left(|R_B(\omega_i,\mu)|^2 - |R_{iB\,exp}|^2 \right)^2 \end{aligned} \right). \tag{4.15}$$

Here K is the number of the measured values of the transmission and reflection coefficients, D_{exp}, $D_{B\ exp}$, R_{exp} and $R_{B\ exp}$ are the measured values of the transmission and reflection coefficients of the electromagnetic wave in the absence of an external magnetic field and when the magnetic field is applied with induction B, respectively.

The required values of the mobility for two measured structures with thicknesses of heavily doped semiconductor layers $t = 2.17$ μm and $t = 13.14$ μm were determined by a numerical method as a result of solving the equation

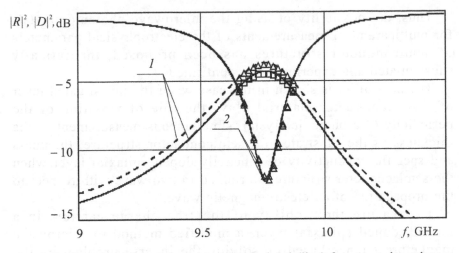

Fig. 4.52. The experimental (points) and calculated (line) frequency dependences of $|D|^2$ (curves *1*) and $|R|^2$ (curves *2*) of the photonic structure (Fig. 4.50) under the action of a magnetic field with induction $B = 0.6$ T (dashed lines) and in its absence (solid lines). The gallium arsenide structure had an epitaxial layer of thickness $t = 13.14$ μm [37].

$$\frac{\partial S(\mu)}{\partial \mu} = 0 \qquad (4.16)$$

and amounted to 0.591 m²/(V·s) and 0.72 m²/(V·s), respectively.

Figure 4.52 shows the experimental (points) and calculated (lines) frequency dependences of the moduli of the reflection and transmission coefficients of the photonic structure (see Fig. 4.50) for a semiconductor structure with an epitaxial layer thickness $t = 13.14$ μm and a mobility value determined from the solution of equation (4.16).

An error estimate of the measurement method for the mobility of the free charge carriers using the photonic structure shown in Fig. 4.50 was carried out using a numerical experiment. The data for the solution of the test problem were set in the form of the values of the transmittance and reflection coefficients of the electromagnetic wave from a photonic crystal with the structure under study containing a gallium arsenide layer of different thicknesses, obtained from expressions (4.15) and (4.16) with an error of ±10%, determined taking into account the results of the full experiment.

The error in the mobility measurement method for structures with an epitaxial layer thickness $t = 2.17$ μm was ±5.0%, with a thickness $t = 13.14$ μm it was ±0.7%.

Thus, the possibility of using the microwave photonic crystals for multiparameter measurements of the electrophysical parameters of semiconductor structures has been proposed, theoretically substantiated and experimentally confirmed.

The possibility is shown in the case when the investigated layer of a semiconductor material plays the role of violation of the periodicity of a photonic crystal, simultaneous measurement of the thickness of the substrate of a semiconductor structure, thickness and specific conductivity of a heavily doped epitaxial layer when the semiconductor structure is oriented in two ways with respect to the propagation of an electromagnetic wave.

To measure the mobility of the free charge carriers in a heavily doped epitaxial layer, a modified method of microwave magnetoresistance based on solving the inverse problem by the method of least squares using the frequency dependences of the transmission and reflection coefficients measured under the action of a magnetic field and in its absence is proposed.

When semiconductor structures, which are n^+–n-structures, are used, it becomes necessary to measure both the parameters of the n^+- and n-layers. A method for simultaneous measurement of three parameters of n^+–n-structures is proposed in [45]: the conductivity of the n-layer, which plays the role of a semiconductor substrate, the thickness and specific conductivity of a heavily doped epitaxial n^+-layer, using a one-dimensional microwave photonic crystal.

The measured structure, placed on the boundary of the violated central teflon layer and the next layer made of Al_2O_3 ceramics, was oriented in two ways with respect to the direction of propagation of the electromagnetic wave. The location of the sample inside the violated layer and its orientation relative to the disturbed layer (configurations 1 and 2) in the photonic crystal are shown in Fig. 4.45 and on the inset to Fig. 4.45.

The investigated samples were epitaxial gallium arsenide structures of thickness $t_S = t_s + t_{sub}$, consisting of a heavily doped epitaxial n^+-layer of thickness t_s with electrical conductivity σ_s and a semiconductor substrate of thickness t_{sub} with electrical conductivity σ_{sub}.

The frequency dependence of the reflection $R(\omega)$ and the transmission $D(\omega)$ coefficients of an electromagnetic wave at its normal incidence on a multilayer structure completely filling the waveguide along the cross section and having layer planes perpendicular to the radiation propagation direction was calculated using the expressions (4.5), (4.6).

For the simultaneous determination of the electrical conductivity σ_{sub} below the semiconductor substrate, the thickness t_s, and the electrical conductivity σ_s of the heavily doped semiconductor n^+-layer, the least squares method was used for the frequency dependences $D(\omega)$ and $R(\omega)$ to determine the values of the parameters σ_{sub}, t_s and σ_s where the sum $S(\sigma_{sub},t_s,\sigma_s)$ of the squares of the differences of the calculated, $|D(\omega,\sigma_{sub},t_s,\sigma_s)|^2$ and $|R(\omega,\sigma_{sub},t_s,\sigma_s)|^2$, and the experimental (initial), $|D_{exp}|^2$ and $|R_{exp}|^2$ values of the squares of the moduli of the transmission and reflection coefficients measured in two different configurations (Fig. 4.45) of a violated photonic crystal

$$S(\sigma_{sub},t_s,\sigma_s) = \sum_{i=1}^{K} \begin{pmatrix} \left(|D_1(\omega_i,\sigma_{sub},t_s,\sigma_s)|^2 - |D_{i1\,exp}|^2\right)^2 + \\[2mm] + \left(|R_1(\omega_i,\sigma_{sub},t_s,\sigma_s)|^2 - |R_{i1\,exp}|^2\right)^2 + \\[2mm] + \left(|D_2(\omega_i,\sigma_{sub},t_s,\sigma_s)|^2 - |D_{i2\,exp}|^2\right)^2 + \\[2mm] + \left(|R_2(\omega_i,\sigma_{sub},t_s,\sigma_s)|^2 - |R_{i2\,exp}|^2\right)^2 \end{pmatrix} \tag{4.17}$$

becomes minimal. Here K is the number of measured values of the transmission and reflection coefficients.

To test the measurement method, a test problem was solved, which consisted of the following: the thicknesses and specific electrical conductivities of the semiconductor layers were determined and the frequency dependences of the transmission and reflection coefficients of the structure studied were calculated using expressions (4.5) and (4.6), that is, a direct problem was solved. These frequency dependences were chosen as the initial ones in solving the inverse problem of finding the parameters of the semiconductor structure, which in this case are considered to be unknown and are to be determined. Comparison of the results of the solution of the inverse problem with the initial values of the thicknesses and the specific electrical conductivities of the semiconductor layers makes it possible to estimate the error of the proposed measurement method.

The required values of the parameters of the investigated sample are determined by the numerical method as a result of solving the system of equations:

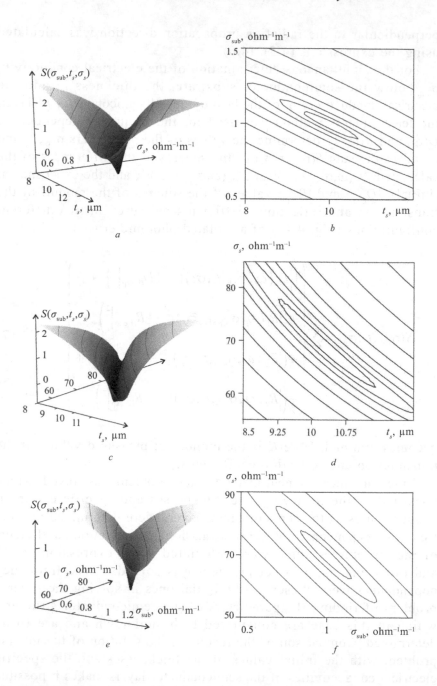

Fig. 4.53. The form of the residual function in space and contour maps in the planes of the required parameters $a, b - (\sigma_{sub}, t_s); c, d - (\sigma_s, t_s); e, f - (\sigma_{sub}, \sigma_s)$ for a sample of gallium arsenide with an epitaxial layer of thickness $t_s = 10.0$ μm and electrical conductivity $\sigma_s = 70.0$ Ohm^{-1}m^{-1} grown on a semiconductor substrate with electrical conductivity $\sigma_{sub} = 1.0$ Ohm^{-1}m^{-1} and a thickness $t_{sub} = 480$ μm [45].

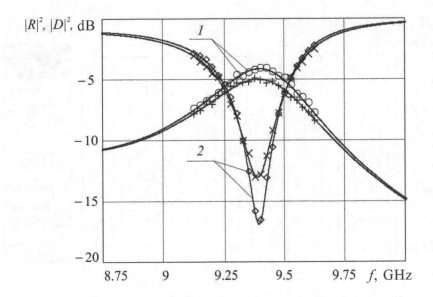

Fig. 4.54. The initial test (points) and calculated (lines), using the results of solving the inverse problem, frequency dependences $|D|^2$ (curves *1*) and $|R|^2$ (curves *2*) for two configurations of the photonic structure (see the inset to Fig. 4.45: configuration 1 – $\Diamond\Diamond\Diamond$ and $\circ\circ\circ$, configuration 2 – xxx and +++) containing a sample of gallium arsenide with an epitaxial layer of thickness t_s = 10.32 μm and electrical conductivity σs = 70.1 ohm^{-1}m^{-1} grown on a high-resistance substrate with a thickness t_{sub} = 480.0 μm and electrical conductivity σ_{sub} = 0.95 ohm^{-1} m^{-1} [45].

$$\frac{\partial S(\sigma_{sub},t_s,\sigma_s)}{\partial \sigma_{sub}}=0 \quad \frac{\partial S(\sigma_{sub},t_s,\sigma_s)}{\partial t_s}=0 \quad \frac{\partial S(\sigma_{sub},t_s,\sigma_s)}{\partial \sigma_s}=0. \quad (4.18)$$

The residual function $S(\sigma_{sub},t_s,\sigma_s)$, determined by the expression (4.17) and represented in Fig. 4.53 *a, c, e*, for the case when the electrical conductivity σ_{sub} under the semiconductor substrate was 1.0 ohm^{-1}m^{-1}, and the thickness and electrical conductivity of the heavily doped GaAs semiconductor n^+-layer were t_s = 10.0 μm and σ_s= 70.0 ohm^{-1}m^{-1}, as follows from the results of the calculation, it has a global minimum in the coordinate space $(t_{sub},\sigma_{sub},\sigma_s,S(t_{sub},\sigma_{sub},\sigma_s))$, and the contour maps (see Figs. 4.53 *b, d, f*) which confirms the possibility of uniquely determining the thickness and electrical conductivity of a semiconductor layer from the solution of the system of equations (4.18).

The values of the specific electrical conductivity of a semi-conductor substrate, the heavily doped epitaxial layer and its

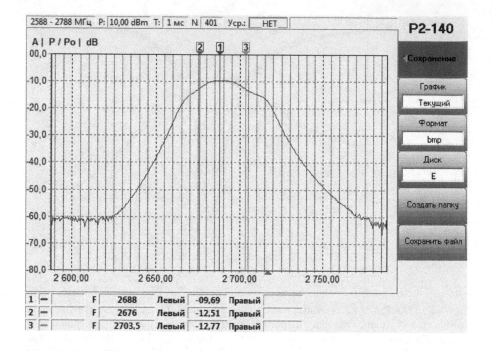

Fig. 4.55. Amplitude–frequency characteristic of the filter based on the band-pass structure [46]

thickness, determined from the solution of the inverse problem using the system of equations (4.18), were $\sigma_s = 0.95$ ohm^{-1}m^{-1}, $t_s = 10.32$ μm and $\sigma_s = 70.1$ ohm^{-1}m^{-1}, respectively.

Figure 4.54 shows the test (initial) and frequency dependences of the moduli of the reflection and transmission coefficients calculated for the two configurations of the photonic structure calculated using the results of solving the inverse problem (see the inset in Fig. 4.45).

The relative error in determining the semiconductor substrate proposed by the method of electrical conductivity, the thickness of the semiconductor layer and its electrical conductivity was ±4.0%, ±0.2%, ±2.6%, respectively.

Thus, for the case when the n^+–n-structure under study plays the role of a violation of the periodicity of a photonic crystal, the results indicate the possibility of simultaneous measurement of the electrical conductivity of a semiconductor substrate, the thickness and specific conductivity of a heavily doped epitaxial layer, with the orientation of a semiconductor structure in two ways with respect to the propagation of an electromagnetic wave.

Another example of the use of the bandwidth ('allowed' band) of the periodic structure in the microstrip arrangement is given by the authors [46] (Fig. 4.55). Such a structure was used in the scheme of frequency multipliers of high multiplicity. When the multiplier is loaded onto such a band-pass structure that plays the role of a filter tuned to the 24th harmonic, the power consumed to excite the remaining harmonics outside the filter bandwidth is reduced. Due to this, most of the power of the input signal input to the multiplier diode is converted to the power of the 24th harmonic.

References

1. Ozbay E., Temelkuran B., Bayindir M., Progress in Electromagnetics Research, 2003. Vol. 41, P. 185–209.
2. Ziroff M. Nalezinski W., in: Proc. of 34th European Microwave Conference. Amsterdam, Netherlands. 12–14th October 2004. Vol. 2. P. 93–96.
3. Usanov D.A., Skripal' A.V., Abramov A.V., et al., Izv. VUZ. Radioelektronika. 2009. Vol. 52, No. 1. P.73 – 80.
4. J. Halszine, Passive and Active Microwave Circuits, John Wiley & Sons, New York, Chichester, Brisbane, Toronto, 1978.
5. Lee K.A., Guo Y., Stimson Ph.A., et al., IEEE Transactions on Antennas and Propagation, vol. 39, No. 3, P. 425-428, 1991.
6. Usanov D.A., Skripal' A.V., Abramov A.V., et al., Measurement of the Metal Nanometer Layer Parameters on Dielectric Substrates using Photonic Crystals based on the Waveguide Structures with Controlled Irregularity in the Microwave Band, Proc. of 37rd European Microwave Conference, 2007, P. 198–201.
7. Usanov D.A., Skripal' A.V., Abramov A.V., et al., Pis'ma Zh. Tekh. Fiz. 2007. Vol. 3, No. 2, P. 13–22.
8. Usanov D.A., Skripal' A.V., Abramov A.V., et al., in: Proc. of 36rd European Microwave Conference. Manchester, UK. 10-15th September 2006. P. 921–924.
9. Chaplygin Yu.A., Usanov D.A., Skripal' A.V., et al., Izv. VUZ. Elektronika, 2006. No. 6. Pp. 27–35.
10. Patent of the Russian Federation for invention No. 2360336. Wide-band waveguide consistent load. Usanov D.A., Skripal' A.V., Abramov A.V., et al., Publ. 27.06.2009. Bul. No. 18. (on application for the invention of 2008106244/09 of 21.02.2008). IPC H01P 7/00 (2006.01)
11. Usanov D.A., Meshchanov V.P., Skripal' A.V., et al., Radiotekhnika. 2015. No. 7. P. 58-63.
12. Usanov D.A., Meshchanov V.P., Skripal' A.V., et al., in: Proceedings of the 25th International Crimean Conference 'Microwave Engineering and Telecommunication Technologies' (KrymiKo'2015). Sevastopol, Crimea, September 6-12, 2015 T. 1, S. 515–516.
13. Patent of the Russian Federation 2 546 578 C2 IPC H01P 1/26 Broadband microstrip matched load. Usanov V.N. Posadsky A.V. Skripal' et al., Bul. 10. Publ. 10/04/2015. Application: 2013137542/08 of 08/09/2013. The patent holder: The Federal State Budget Educational Institution of Higher Professional Education 'Saratov State University named after N.G. Chernyshevsky.'

14. Usanov D.A., Skripal' A.V., Abramov A.V., Bogolyubov A.S., Zh. Tekh. Fiz. 2006. Vol. 76, No. 5. P. 112–117.

15. Usanov D.A., Skripal' A.V., Abramov A.V., Bogolyubov A.S., et al., in: Proc. of 36rd European Microwave Conference. Manchester, UK. 10-15th September 2006. 509–512.

16. Usanov D.A., Skripal' A.V., Abramov A.V., Bogolyubov A.S., et al., Izv. VUZ. Elektronica. 2007. No. 6. Pp. 25–32.

17. Anlage S.M., Steinhauer D.E., Feenstra B. J., et al., Near-field microwave microscopy of materials properties, in: Microwave Superconductivity. Eds. H. Weinstock and M. Nisenoff. Amsterdam. The Netherlands: Kluwer, 2001. P. 239–269.

18. Usanov D.A., Nearfield scanning Microwave Microscopy and its applications. Saratov: Sarat Publishing House. University, 2010. 100 s.

19. Usanov D.A., Gorbatov S.S.. Near-field effects in electrodynamic systems with inhomogeneities and their use in microwave technology. Saratov: Publishing house Sarat. Univ., 2011.

20. Usanov D.A., Gorbatov S.S.. Izv. VUZ. Radiofizika. 2001. P. 44, No. 12. P. 1046–1049.

21. Usanov D.A., Gorbatov S.S., Izv. VUZ. Radioelektronika. 2002. Vol. 45, No. 9. P. 26–28.

22. Kleismit R. A., Kazimierczuk M.K., Kozlowski G., IEEE Transactions on Microwave Theory and Techniques. 2006. Vol. 54, No. 2. P. 639–647.

23. Usanov D.A., Vagarin A.Yu., Bezmenov A.A., Device for measuring dielectric permittivity of materials. The author's certificate of the USSR No. 1114979. Priority from 22.06.82. Pub. 08/07/84. Bul. No. 35.

24. Usanov D.A., Gorbatov S.S., et al., Pis'ma Zh. Tekh. Fiz.. 2000. Vol. 26, No. 18. P. 47–49.

25. Usanov D.A., Gorbatov S.S., Izv. VUZ. Radioelektronika. 2006. Vol. 49, No. 2. P.27–33.

26. Usanov D.A., Gorbatov S.S., Pribory tekhnika eksperimenta. 2003. No. 1. P. 72–73.

27. Device for measuring the parameters of materials: Pat. 2373545 RF, Usanov D.A., et al., Saratov University. Publ. 20.11.2009. Bul. No. 32. Priority of 3.06.2008.

28. Usanov D.A., Nikitov S.A. Skripal' A.V., Frolov A.P., Radiotekhnika i elektronika. 2013, Vol. 58. № 12. P. 1071–1078.

29. The patent of the Russian Federation for useful model 144 869 U1 МПК The device for definition of dielectric permeability of plates and thicknesses of nanometer conductive films / D.A. Usanov, S.A. Nikitov, A.V. Skripal', V.E. Orlov, A.P. Frolov. G01N 22/00 B82B 1/00 Application: 2013125178/07 dated 05/30/2013; Opubl. 10/09/2014; Bul. 25.

30. Usanov D.A., Orlov V.E,. Bezmenov A.A., Elektronnaya tekhnika. Elektronika SVCh. 1977, Vol.1, No. 3, P. 37–41.

31. Usanov D.A., Skripal' A.V., Abramov A.V., et al., Izv. VUZ. Elektronika. 2008. No. 5. P. 25–32.

32. Usanov D.A., et al., in: Proc. of European Microwave Week 2008: 38th European Microwave Conference. 27-31 October 2008. Amsterdam, The Netherlands. P. 785–788.

33. Usanov D.A., Skripal' A.V., Abramov A.V., et al., Zh. Tekh. Fiz. 2010. V. 80, No. 8, pp. 143–148.

34. Patent of the Russian Federation 2419099 IPC G 01 R 27/26. A device for measuring the relative permittivity and tangent of the dielectric loss angle of a fluid / Usanov D.A., Skripal' A.V., Abramov A.V., et al. 05/20/2011. Bul. 14.

35. Maloratsky L.G., Yavich L.R., Design and calculation of microwave elements on

strip lines. Moscow: Sov. radio. 1972.

36. Sato N., Buchner R., J. Phys. Chem. A. 2004. Vol. 108, P. 5007–5015.

37. Usanov D.A., Nikitov S.A., Skripal' A.V., Radiotekhnika i Elektronika, 2016, Vol. 61, No. 1, p. 45–53

38. Usanov D.A., Microwave methods for measuring semiconductor parameters. Saratov: Saratov University Publishing House. 1985.

39. Chaplygin Yu.A., Usanov D.A., Skripal' A.V., Izv. VUZ. Elektronika. 005, No. 1, P. 68.

40. Molnar B., Kenedy T.A., J. Electrochem. Soc. Solid-State Science and Technology, 1978, Vol. 125, No. 8, P. 1318.

41. Bannikov V.S., Kachurovsky Yu.G., Petrenko I.V., et al. Elektronnaya promyshlennost', 1982, No. 9, P.48.

42. Bezruchko S.M., Podshivalov V.N., Fisun A.I., Elektronnaya promyshlennost', 1986, No. 3, p. 66.

43. Venig S.B., Usanov D.A., Soloukhin N.G.,, Bragin S.M., Elektronnaya tekhnika. Ser. 8, 1990, Vol. 1, p. 64.

44. Usanov D.A., Skripal' A.V., Abramov A.V., Pozdnyakov V.A, Izv. VUZ. Elektronika. 2004, No. 2, P. 76.

45. Usanov D.A., Nikitov S.A., Skripal' A.V., et al., Semiconductors. 2016. Vol. 50, No. 13, P. 1759–1763.

46. Usanov D.A., Posadsky V.N., Skripal A.V., et al., Izv. VUZ Rossii. Radioelektronika. 2014. No. 4. pp. 48–50.

Conclusion

The results of theoretical and experimental studies of the properties of one-dimensional microwave photonic crystals described in this monograph show that it is possible to achieved their good quantitative agreement and, on this basis, to measure the physical parameters of metal–dielectric and semiconductor structures that play the role of a violation of periodicity.

Various types of photonic crystals are considered: on rectangular waveguides with dielectric inserts, on plane transmission lines, on waveguide-slot lines, on waveguide resonant diaphragms.

The possibility of efficient control of the amplitude–frequency characteristics of microwave photonic crystals is shown.

A new type of microwave photonic crystals is proposed – low-dimensional microwave photonic crystals, and the possibility of the existence of a defect mode in them is shown. Examples of new applications of the properties of the photonic crystals are given: to create matched microwave loads, microwave filters intended for use in circuits with frequency multipliers.

The development of research in this direction can lead to the creation of new types of microwave devices that have improved basic and sometimes unique characteristics.

Index